AF845300

Recovery of *Cryptosporidium* Oocysts From High-Volume Water Samples

The mission of the Awwa Research Foundation (AwwaRF) is to advance the science of water to improve the quality of life. Funded primarily through annual subscription payments from over 1,000 utilities, consulting firms, and manufacturers in North America and abroad, AwwaRF sponsors research on all aspects of drinking water, including supply and resources, treatment, monitoring and analysis, distribution, management, and health effects.

From its headquarters in Denver, Colorado, the AwwaRF staff directs and supports the efforts of over 500 volunteers, who are the heart of the research program. These volunteers, serving on various boards and committees, use their expertise to select and monitor research studies to benefit the entire drinking water community.

Research findings are disseminated through a number of technology transfer activities, including research reports, conferences, videotape summaries, and periodicals.

Recovery of *Cryptosporidium* Oocysts From High-Volume Water Samples

Prepared by:
Jennifer L. Clancy, **Randi M. McCuin**, and **Thomas M. Hargy**
Clancy Environmental Consultants, Inc.
PO Box 314, St. Albans, VT 05478

Sponsored by:
Awwa Research Foundation
6666 West Quincy Avenue, Denver, CO 80235-3098

Published by:
Awwa Research Foundation
Advancing the Science of Water®
IWA Publishing

DISCLAIMER

This study was funded by the Awwa Research Foundation (AwwaRF). AwwaRF assumes no responsibility for the content of the research study reported in this publication or for the opinions or statements of fact expressed in the report. The mention of trade names for commercial products does not represent or imply the approval or endorsement of AwwaRF. This report is presented solely for informational purposes.

Published by IWA Publishing, Alliance House, 12 Caxton Street, London SW1H 0QS, UK

Telephone: +44 (0) 20 7654 5500; Fax: +44 (0) 20 7654 5555; Email: publications@iwap.co.uk;
Web: **www.iwapublishing.com**

AwwaRF report number 90960F. Originally published by AwwaRF for its subscribers in 2003.
IWA Publishing version published 2004.

Copyright © 2003
by
Awwa Research Foundation

ISBN 1 84339 867 2

TABLE OF CONTENTS

TABLES

FIGURES

FOREWORD

The Awwa Research Foundation is a nonprofit corporation that is dedicated to the implementation of a research effort to help utilities respond to regulatory requirements and traditional high-priority concerns of the industry. The research agenda is developed through a process of consultation with subscribers and drinking water professionals. Under the umbrella of a Strategic Research Plan, the Research Advisory Council prioritizes the suggested projects based upon current and future needs, applicability, and past work; the recommendations are forwarded to the Board of Trustees for final selection. The foundation also sponsors research projects through the unsolicited proposal process; the Collaborative Research, Research Applications, and Tailored Collaboration programs; and various joint research efforts with organizations such as the U.S. Environmental Protection Agency, the U.S. Bureau of Reclamation, and the Association of California Water Agencies.

This publication is a result of one of those sponsored studies, and it is hoped that its findings will be applied in communities throughout the world. The following report serves not only as a means of communicating the results of the water industry's centralized research program but also as a tool to enlist the further support of the nonmember utilities and individuals.

Projects are managed closely from their inception to the final report by the foundation's staff and large cadre of volunteers who willingly contribute their time and expertise. The foundation serves a planning and management function and awards contracts to other institutions such as water utilities, universities, and engineering firms. The funding for this research effort comes primarily from the Subscription Program, through which water utilities subscribe to the research program and make an annual payment proportionate to the volume of water they deliver and consultants and manufacturers subscribe based on their annual billings. The program offers a cost-effective and fair method for funding research in the public interest.

A broad spectrum of water supply issues is addressed by the foundation's research agenda: resources, treatment and operations, distribution and storage, water quality and analysis, toxicology, economics, and management. The ultimate purpose of the coordinated effort is to assist water suppliers to provide the highest possible quality of water economically and reliably. The true benefits are realized when the results are implemented at the utility level. The foundation's trustees are pleased to offer this publication as a contribution toward that end.

This study is of importance to the water industry because there is great interest among water utilities, regulators, and consumers about the occurrence of *Cryptosporidium* oocysts in water supplies. The methodologies available for recovery and detection of the parasite had been improved over the years, but the methods focused on collection and analysis of relatively low sample volumes of source water; there was no usable finished drinking water method. This project improved upon the existing methods for oocyst analysis of water permitting larger volumes of source water to be collected and analyzed, and developed a method for 1000 L finished drinking water samples. All methods developed in the project were validated using the criteria specified by the USEPA, so that the methods can be used for compliance monitoring. These new methods are now in widespread use in the U.S. and in other countries.

Edmund G. Archuleta, P.E.
Chair, Board of Trustees
Awwa Research Foundation

James F. Manwaring, P.E.
Executive Director
Awwa Research Foundation

ACKNOWLEDGMENTS

The help and guidance of the project managers, Kate Martin and Emerson Lomaquahu, and the AwwaRF Project Advisory Committee were greatly appreciated. The PAC members were: Lisa Blancero (New York City Department of Environmental Protection), John Troutt (Tampa Bay Water), and William Yanko (private consultant). The authors gratefully acknowledge the support of water utilities that provided samples for testing:

Champlain Water District, South Burlington, VT
City of St. Albans, St. Albans, VT

The authors also wish to thank Tricia Klonicki of CH Diagnostics and Consulting Services, Inc. (Loveland, CO), Dr. Fu-Chih Hsu of UL-Environmental Health Laboratories (South Bend, IN), Lisa Blancero of the New York City Department of Environmental Protection (NY), and Maurie Simpson of the Metropolitan Water District of Southern California (LaVerne, CA) for participating in the round robin studies.

EXECUTIVE SUMMARY

After two decades, waterborne outbreaks of cryptosporidiosis are still a source of major public health concern. Numerous waterborne outbreaks of human disease have occurred in Canada, Japan, the United Kingdom (U.K.) and United States (U.S.). To be able to better control waterborne disease, accurate oocyst occurrence information is needed for water sources and treated supplies, and these data can be of value only when sufficiently sensitive methods are employed. Although numerous surveys have been conducted in various geographic locations, the methods used for testing were often inefficient and lacked sensitivity. The U.S. Environmental Protection Agency (USEPA) has improved methods for analysis of *Cryptosporidium* and *Giardia* in water. However, these methods employ small sample volumes (10 L) and were designed for raw water analysis. The occurrence of oocysts in finished water is likely to be low and sporadic, requiring methods that permit higher sample volumes and increased sensitivity. The industry recognized that improved methods for *Cryptosporidium* analysis of water were needed, and this project undertook the development of these improved methods.

RESEARCH OBJECTIVES

The Awwa Research Foundation (AwwaRF) funded this study to develop methods for high volume sampling of source and treated water, and to collaboratively test the new methods to validate them as USEPA standards for *Cryptosporidium* analysis of water.

The specific objectives of this research were:

1) Evaluate three types of filters approved for use in other standards in the U.S. or U.K. and which show promise for being adapted to higher volumes.
2) Develop one or more methods for concentration of sample eluants so that centrifugation of the eluant can be avoided.
3) Examine immunomagnetic separation (IMS) systems for their abilities to handle high volume source water and treated water concentrates.
4) Collaboratively test at least one high volume method using the performance based measurements system (PBMS) criteria so that these methods are USEPA-approved methods at the conclusion of the study.

APPROACH

The approach undertaken was to use USEPA Method 1622: *Cryptosporidium* in Water by Filtration/IMS/FA as the basic method upon which improvements would be made. Method 1622 is a robust method and is the currently accepted USEPA standard for oocyst recovery and enumeration from water (USEPA, 2001a). Many other countries have adopted Method 1622 as their standard for *Cryptosporidium* testing. This study examined individual steps in the methodology and sought to improve and then optimize these steps. After optimization, the steps were combined into a full method for single laboratory validation. The final methods were collaboratively tested in four other laboratories to develop performance criteria required under the USEPA PBMS format.

CONCLUSIONS

1. The filters validated in Method 1622 could be used after modifications in the protocols for use with high volume source water samples.
2. A new filter, the Gelman Envirochek HV, was tested, optimized and Tier 2 validated for use with increased (50 L) source water samples.
3. The IDEXX Filta-Max system was Tier 2 validated for use in Method 1622.
4. An alternative elution/concentration protocol for the Filta-Max foam module was developed using the stomacher and centrifugation, and Tier 1 validation was completed in three labs. This technique reduces hands-on time to elute the filter and minimizes clean up between samples.
5. Two high volume finished water methods were developed and single laboratory-tested.
6. A method for 1000 L finished drinking water analysis was developed and Tier 1 validated using the Envirochek HV with a new elution protocol employing sodium hexametaphosphate.
7. The IDEXX FiltaMax foam cell filter was shown to permit equivalent *Cryptosporidium* recovery in 1000 L tap water samples using Method 1622.
8. A new IMS kit developed by Aureon Biosystems underwent Tier 1 validation for Method 1622.
9. The Aureon IMS kit was shown to be equivalent to the Dynal IMS kit for use with 1000 L finished drinking water samples.

RECOMMENDATIONS

The methods developed and validated in this research permit collection and analysis of significantly higher sample volumes with no decrease in method sensitivity. They are robust as was demonstrated in collaborative testing in four other laboratories. However, additional work remains.

1. The high volume finished drinking water methods were not collaboratively tested in this study but should undergo round robin testing as soon as EPA develops criteria for validation.
2. Molecular methods are available and need to be incorporated into these methods to allow determination of species and viability/infectivity status so that data can be used for risk assessment and public health decision-making.

FUTURE RESEARCH

New technologies are needed to improve upon the *Cryptosporidium* analytical methods. The advances noted between the Information Collection Rule (ICR) Method and Method 1622 were due to improved filtration and the use of IMS. These techniques have been fully developed and maximally optimized for recovery of *Cryptosporidium* in water; new technologies are needed to improve methods beyond the current status. From sample collection to analysis, the process is still labor intensive, expensive, incomplete as far as species and viability or infectivity are concerned, and is not available for real-time decision-making. Combinations of new molecular techniques or on-line monitoring capabilities will be needed to truly advance the science of *Cryptosporidium* monitoring of water supplies.

CHAPTER 1
INTRODUCTION

BACKGROUND

Since the late 1980's, water suppliers have conducted protozoan monitoring on a voluntary basis to develop baseline data on occurrence, evaluate treatment processes, design new treatment, and support investment planning (Clancy and Hansen, 1999). In the U.S., the USEPA ICR method had been used for producing the bulk of these data (USEPA 1996), but with the introduction of USEPA Method 1622: *Cryptosporidium* in Water by Filtration/IMS/FA (USEPA, 2001a) and Method 1623: *Giardia* and *Cryptosporidium* in Water by Filtration/IMS/FA (USEPA, 2001b), the ICR Method was replaced. These new methods differ from the ICR Method in that:

1) sample collection filters permit 100% capture of (oo)cysts,
2) subsequent elution provides ≥80% removal of target organisms from the filters,
3) the entire packed pellet volume can be separated and analyzed, eliminating sub-sample analysis, and
4) the addition of *4'-6'-diamidino-2-phenylindole* (DAPI) staining permits easier confirmation of putative (oo)cysts.

The methods are more robust and have greater sensitivity, allowing better data to be developed. The USEPA conducted additional monitoring after the ICR to supplement the ICR database, and these surveys, using Methods 1622 and 1623 were known as the ICR Supplemental Surveys (ICRSS) (USEPA, 2001c). Connell et al (2000) found that the data from the Method 1622 and 1623 interlaboratory validation studies were accurate predictors of method performance during real world monitoring during the ICRSS. Protozoa occurrence data from the 40 water treatment plants that participated in both the ICR and the ICRSS indicated that Methods 1622 and 1623 results generated under the ICRSS were characterized by higher detection rates than the results generated by the ICR protozoa method, with fewer potential false positive results.

While methods 1622 and 1623 are the best methods available for protozoan monitoring, they are limited in that they pertain to source water testing and are limited to 10 L sample volumes. For some source waters, particularly those of low turbidity, sample volumes could be increased to 50 L or more. For finished drinking water, a 10 L sample volume is too low to provide any useful information on possible contamination. The need for higher sample volumes of finished drinking water has been an important gap in method development for years. The focus in the U.S. has been on source water testing, for the ICR and ICRSS, and most recently for the upcoming Long Term 2 Enhanced Surface Water Treatment Rule (LT2ESWTR). Finished drinking water testing is done, but using the ICR method which is recognized to be poor (Clancy et al, 1994; Clancy et al, 1999; Clancy and Hansen, 1999).

The immunomagnetic separation (IMS) step, an important development in protozoa analysis of water, had not been optimized in Method 1622 beyond the manufacturer's recommendation, although in single laboratory studies it was found that the packed pellet volume could be doubled without compromising recoveries (Bukhari et al, 1998; Rochelle et al, 1999). Refinements in the use of the IMS technology were another factor to be evaluated in improvement of methods for high volume analysis.

The primary goal of this study was to develop ways to alter Method 1622 so that higher volumes of source water (up to 50 L) and 1000 L of finished drinking water could be analyzed for *Cryptosporidium*, maintaining the method robustness characteristic of Method 1622. Once developed, any new method(s) must undergo collaborative testing to determine its performance characteristics. The research team conducted the collaborative testing using the USEPA's PBMS criteria for method equivalency.

APPROACH

The approach undertaken was to use USEPA Method 1622: *Cryptosporidium* in Water by Filtration/IMS/FA as the basic method upon which improvements would be made. Method 1622 is a robust method and is the currently accepted USEPA standard for oocyst recovery and enumeration from water (USEPA, 2001a). Many other countries have adopted Method 1622 as their standard for *Cryptosporidium* testing. This study examined individual steps in the methodology and sought to improve and then optimize these steps. After optimization, the steps were combined into a full method for single laboratory validation. The final methods were collaboratively tested in other laboratories to develop performance criteria under the USEPA PBMS format.

SAMPLING

Sample collection is the first critical component of the protozoan method and can affect overall recovery significantly. The target organisms must be effectively captured and retained, and then be able to be released from the capture system for subsequent detection and enumeration. This study used the three Method 1622-approved sampling filters, the Envirochek capsule, the Filta-Max foam cell filter, and the CrypTest cartridge filter. A new filter from Pall Gelman, the Envirochek HV capsule, was also tested and optimized in this study.

Pall Gelman Envirochek Capsule

The standard Pall Gelman Envirochek capsule is a membrane device for the simultaneous concentration of *Cryptosporidium* oocysts and *Giardia* cysts from water (Matheson et al, 1998). Samples are filtered through a Supor® polyethersulfone membrane with a 1.0 μm pore size that is enclosed in a plastic capsule. (Oo)cysts are mechanically eluted from the membrane fiber using a wrist action shaker and a non-ionic detergent, and concentrated by centrifugation. The concentrate can be further processed using any separation technique to separate the target organisms from other debris. The Envirochek capsule was developed for use in Methods 1622 and 1623 with 10 L sample volumes. The Envirochek HV capsule was developed specifically for use with increased sample volumes. The Envirochek HV is identical to the standard Envirochek capsule except that the pleated filter in the capsule is a polyester track-etch membrane material in place of polyethersulfone.

Whatman CrypTest Cartridge

The Whatman CrypTest cartridge filter is the former Corning CrypTest filter. The filter material, polycarbonate track etch (PCTE) membrane, is made from polymer films. These films have been subjected to bombardment by high-energy ions, creating tracks of damage along their course of travel. The membrane is then chemically etched for a prescribed time to create precisely sized pores along these tracks. The filtering mechanism is a sieving or screening one, where particles larger than the pores do not pass. The CrypTest cartridge uses a pleated format membrane to increase surface area for filtration, coupled with modifications to the elution process that minimize handling and improve recoveries through the use of a backwashing technique. The CrypTest cartridge is designed for routine protozoan sampling and can be used with variable volumes of raw or treated water. This filter meets the criteria established by the USEPA for the recovery of (oo)cysts by the filtration step in Methods 1622 and 1623 and was validated under the performance based measurements system criteria for Method 1623.

IDEXX Filta-Max Foam Filter

The IDEXX Filta-Max foam filter is comprised of 60 open cell reticulated foam pad rings that when compressed produce an effective pore size of ~1 μm. The FiltaMax system is a complete sample collection, filter elution, and concentration system initially designed in the U.K. by Genera Technologies. This system was the original one approved by the U.K. Drinking Water Inspectorate for use in their finished water continuous monitoring regulation (Anon, 1989). In the U.S., the Filta-Max is an approved method for sample volumes up to 50 L for Methods 1622 and 1623.

IMMUNOMAGNETIC SEPARATION

Occurrence of *Cryptosporidium* oocysts in low numbers in environmental water necessitates filtration of numerous liters of sample through a filter. The filter is eluted and subsequent concentration is done using both large and small-scale centrifugation. The concentration process leads to accumulation of debris in the sample and in turn, detection of organisms requires clarification processes to separate organisms from debris. The earlier flotation procedures for recovery of oocysts from background debris were highly variable, and have been replaced by immunomagnetic separation.

At the outset of this study, the Dynal Dynabeads (Dynal A.S., Oslo, Norway) IMS kits were the only ones commercially available; the introduction near the end of the study of the Aureon BioSystems (Vienna, Austria) IMS kit permitted evaluation and validation of a second IMS kit for use in Method 1622 and for high volume finished drinking water analysis.

The two commercially available IMS kits use the same concept for recovering target organisms and utilize a *Cryptosporidium*-specific monoclonal antibody (mAb) which is covalently linked to magnetic beads (bead conjugate) of a pre-defined size (0.8 to 4.5 μm). Mixing the bead conjugate in a buffered sample, for an optimized time period, enables the mAb to capture target organisms. Application of a magnetic field enables separation of the bead conjugate-organism complex from the remainder of the sample. The bead conjugate-organism complex, which is formed as a result mAb binding to epitopes expressed on the outer wall of the organism, can then be suspended in an acidic solution. With vigorous agitation, the bead conjugate can be dissociated from the antigens captured on the surface of target organisms. As mAb reactivity is pH dependent, the sample concentrate requires neutralization prior to detection/enumeration and this can be performed on well slides in a format consistent with the staining procedures described in Method 1622.

SAMPLE MATRIX

Methods 1622 and 1623 were developed and validated originally for 10 L source water samples. When higher volumes were used with the standard Envirochek capsule, the increased volumes could be passed through the filter, but the overall recoveries were significantly diminished (Clancy et al, 2000). This was not noted with the Filta-Max or CrypTest filters. This observation led Pall Gelman to develop the Envirochek HV, designed for sampling increased volumes of source water (50 L) and high volumes (1000 L) of treated drinking water.

This project focused a great deal of attention on developing a finished drinking water method since the only method in the U.S. for finished drinking water testing was the ICR method, which is characterized by poor performance (Clancy et al, 1994; Clancy et al, 1999). The IDEXX Filta-Max is used in the U.K. for 1000 L finished drinking water sampling and experiments repeated in this study showed it performs well, yielding high and reproducible recoveries. The Filta-Max was also validated for use in Method 1623 with 50 L source water samples.

Adapting the Envirochek HV for high volume finished water samples proved to be more of a challenge and required an alteration in the buffer composition to achieve high and reproducible recoveries. While capture on the filters was complete, elution of the target organisms proved difficult and recoveries were low. When 1000 L samples were spiked with 100 to 500 cysts and oocysts, recoveries were often in the single digits, in stark contrast to the performance of the standard Envirochek or Envirochek HV in source water matrices. It was hypothesized that chemistry of finished drinking water, either from the treatment process itself, or from the distribution system, was interfering with the elution and/or separation processes. Laboratories in the U.K. and Australia reported similar isolated findings. It was also recognized that when using the standard elution procedures for these filters following filtration of a high volume (500 to 1000 L) of finished water, residual material remained on the filter surfaces, and that extraordinary means would be necessary to complete the elution. The answer was the inclusion of sodium hexametaphosphate in the elution buffer. Sodium hexametaphosphate has been used for many years to reduce scaling (Fireman and Reitemeier, 1944), and is the main ingredient used in household products to aid cleansing in hard water conditions. In drinking water applications, it is used as a scale inhibitor in membrane pre-treatment (Schock, 1990; Kasper, 1993).

PERFORMANCE BASED MEASUREMENTS SYSTEM (PBMS)

Background

The USEPA Office of Water developed a set of criteria for approval of water methods that is described in 40 CFR Parts 136 and 141 (USEPA, 1997). The system is known as the performance based measurements system (PBMS) under the Clean Water Act and Safe Drinking Water Act for compliance monitoring programs. PBMS is designed to expand flexibility and approval of new methods. USEPA's stated six goals for PBMS are:

- Provide a simple, straightforward way for the regulatory community to respond to specific measurement needs with reliable, cost-effective methods.
- Emphasize project- or application-specific method performance needs rather than requiring that specific measurement technologies be used in order to avoid costly measurement overkill.
- Encourage the laboratory community to use professional judgment in modifying or developing alternatives to established agency methods.
- Employ a consistent way to express method performance criteria that is independent of the type of method or technology used, including articulating measurement needs in qualitative and quantitative terms.
- Foster new technology development and continuous improvement in measurement methodology, by providing qualitative and quantitative targets for identification measurement gaps to method developers and other researchers.
- Encourage the measurement community to give the agency feedback on new monitoring approach successes as well as failures in order to expand our knowledge of new or modified approaches and to assist others by helping to disseminate this information to the wider monitoring community.

PBMS expands flexibility to modify methods and provides a mechanism for non-EPA organizations to develop new methods for water analysis (USEPA, 1997; 2001a). Acceptance criteria are set for reference methods through collaborative testing of the methods in experienced laboratories. Individual laboratories are then able to introduce modifications to these methods as long as any method modification produces data of equivalent or superior quality to the data produced by the approved method that has been modified. This means that alternate techniques or materials that are more cost effective and/or allow the generation of higher quality data can quickly be introduced through the testing requirements of PBMS.

Methods 1622 and 1623 were the first microbiological test methods to be developed incorporating PBMS. Both methods contain extensive quality assurance/quality control (QA/QC) requirements as part of the method. These include successful demonstrations of both initial precision and recovery (IPR) and matrix spikes (MS). IPR is the ability of a lab to generate data that fall within the method acceptance criteria in reagent water samples. Reagent water is defined here as a water matrix that is devoid of the analyte of interest, and is usually deionized (DI) or filtered tap water. MS samples are those that are actual source water matrices. In both cases the samples are spiked with *Cryptosporidium* (Method 1622) or *Giardia* and *Cryptosporidium* (Method 1623), and the sample is processed according to the method. Unspiked samples are also processed as negative controls for IPR and to determine background levels of (oo)cysts for MS samples. A third category of samples is the ongoing precision and recovery (OPR). These are the spiked reagent water samples that are required weekly as part of the method QC. EPA has developed acceptance criteria for Methods 1622 and 1623 based on the results from 293 *Cryptosporidium* OPR samples and 186 *Giardia* OPR samples analyzed by six laboratories during the ICRSS (Connell et al, 2000; USEPA, 2001d, 2001e). The acceptance criteria are shown in the following tables.

Table 1.1
Quality Control Acceptance Criteria for *Cryptosporidium* Using Method 1622[1]

Quality Control Acceptance Criteria for *Cryptosporidium* Using Method 1622[1]

Performance Test	Section[2]	Acceptance Criteria
Initial precision and recovery	9.4	
Mean recovery	9.4.2	24-100%
Precision (as maximum relative standard deviation)	9.4.2	55
Ongoing precision and recovery	9.7	11-100%
Matrix/matrix spike duplicate (for method modifications)	9.5	
Mean recovery[3]	9.5.2	13-143%
Precision (as maximum relative percent difference)	9.5.2	67

[1] USEPA Method 1622: *Cryptosporidium* in Water by Filtration/IMS/FA (USEPA, 2001a)
[2] Refer to specific section in Method 1622
[3] The acceptance criteria for mean MS/MSD recovery serves as the acceptance criteria for MS recovery during routine use of the method (Section 9.5.1).

Table 1.2.
Quality Control Acceptance Criteria for *Cryptosporidium* and *Giardia* Using Method 1623[1]

Quality Control Acceptance Criteria for *Cryptosporidium* and *Giardia* Using Method 1623[1]

Performance Test	Section[2]	Acceptance Criteria	
		Cryptosporidium	*Giardia*
Initial precision and recovery	9.4		
Mean recovery	9.4.2	24-100%	24-100%
Precision (as maximum relative standard deviation)	9.4.2	55	49
Ongoing precision and recovery	9.7	11-100%	14-100%
Matrix/matrix spike duplicate (for method modifications)	9.5		
Mean recovery[3]	9.5.2	13-111%	15-118%
Precision (as maximum relative percent difference)	9.5.2	61	30

[1] USEPA Method 1623: *Giardia* and *Cryptosporidium* in Water by Filtration/IMS/FA (USEPA, 2001b)
[2] Refer to section in Method 1623
[3] The acceptance criteria for mean MS/MSD recovery serves as the acceptance criteria for MS recovery during routine use of the method (Section 9.5.1).

Using PBMS To Modify Methods 1622 And 1623

There are two ways described in Methods 1622 and 1623 for laboratories to introduce method modifications to improve performance or efficiency. Tier 1 validation is used when an individual laboratory wishes to alter a component or step in the method for in-house use. A Tier 1 validation includes analysis of four replicates of spiked reagent water and a method blank, and is equivalent to a demonstration of IPR. If recoveries fall within the acceptance criteria for both recovery and precision shown in Tables 1.1 and 1.2, that laboratory can adopt the method modification. Matrix spikes are still required on each water type to which the modification will be applied and on every 20th sample of that water thereafter as described in the method QA/QC.

Tier 2 validation expands Tier 1 to a collaborative or multi-laboratory study, and includes a minimum of three laboratories that must generate data that meet the method acceptance criteria. If more than three labs participate in the study, 75% of those labs must meet the acceptance criteria. Tier 2 differs from Tier 1 in that the analysis includes a set of IPR samples AND a set of matrix spike/matrix spike duplicates (MS/MSD) samples and an unspiked matrix sample. Tier 2 is Tier 1 plus a set of MS/MSD samples analyzed by at least three laboratories. Acceptance criteria from each lab must fall within the percent recovery and precision ranges for both the IPR and MS/MSD for all three labs (or 75% of the participating labs) for a successful Tier 2 validation. Modifications that successfully undergo Tier 2 validation are reviewed by EPA and will appear in published versions of the method, that is, as the method is updated and republished by EPA from time to time, the Tier 2 validated modifications become part of the standard method. The details of Tier 1 and 2 validations are summarized in Table 1.3.

Table 1.3
Tier 1 and Tier 2 Validation/Equivalency Demonstration Requirements

Test	Description	Tier 1 Modification[1]	Tier 2 Modification[2]
IPR (Section 9.4)	4 replicates of spiked reagent water	Required. Must be accompanied by a method blank	Required per laboratory
Method blank (Section 9.6)	Unspiked reagent water	Required	Required per laboratory
MS (Section 9.5.1)	Spiked matrix water	Required on each water to which the modification will be applied and on every 20th sample of that water thereafter. Must be accompanied by an unspiked field sample collected at the same time as the MS sample	Not required
MS/MSD (Section 9.5)	2 replicates of spiked matrix water	Recommended, but not required. Must be accompanied by an unspiked field sample collected at the same time as the MS sample	Required per laboratory. Each laboratory must analyze a different water

From USEPA Method 1623: *Giardia* and *Cryptosporidium* in Water by Filtration/IMS/FA (USEPA, 2001b)

[1] If a modification will be used only in one laboratory, these tests must be performed and the results must meet all of the QC acceptance criteria in the method (these tests are also required the first time a laboratory uses the validates version of the method).

[2] If nationwide approval of a modification is sought for one type of water matrix, (such as surface water), a minimum of 3 laboratories must perform the tests and the results of each laboratory individually must meet al of the QC acceptance criteria in the method. If more than 3 laboratories are used in a study, a minimum of 75% of the laboratories must meet all QC acceptance criteria.

Successful Tier 2 validations include:

1) Whatman CrypTest cartridge filter for Methods 1622 and 1623,
2) IDEXX Filta-Max for 50 L source water samples in Methods 1622 and 1623,
3) Pall Gelman Envirochek HV for 50 L source water samples for Method 1622
4) Waterborne, Inc. Aqua-Glo mAbs for *Giardia* and *Cryptosporidium* detection in Methods 1622 and 1623.

CHAPTER 2
MATERIALS AND METHODS

SOURCES OF OOCYSTS

Source of *C. parvum* Oocysts

The oocysts used in this research were from the Iowa isolate maintained at the Sterling Parasitology Laboratory in Tucson, AZ and were originally obtained from Harley Moon (National Animal Disease Center, Ames, IA). This isolate was selected because it is well characterized and has been used in numerous studies including restriction fragment length polymorphism analysis where two separate passages through calves demonstrated identical banding patterns (Ortega et al, 1991). The Iowa isolate has been used to determine the *C. parvum* dose response in neonatal mice (Korich et al, 1990; Finch et al, 1993; Korich et al, 2000). More recently, this isolate has been used in studies examining infection characteristics and immunology in controlled experiments with healthy human volunteers (Dupont et al, 1995; Chappell et al, 1996; Okhuysen et al, 1998; Okhuysen et al, 1999)

Oocysts were produced by feeding 2 x 10^8 oocysts suspended in sterile water to Holstein calves within 4 to 12 hours (h) of birth. This was followed 1 to 2 h later by four pints of colostrum and prophylactic doses of oral Rota/Corona Virus and *E. coli* vaccines. The calves were maintained on a combination of milk replacer, non-fat milk powder, and electrolyte mix that provided 2200 Cal/day. The University of Arizona Institutional Animal Care and Use Committee approved all protocols for calf care and infection.

The total daily fecal output from the infected calves was collected, screened through sieves, and concentrated by centrifugation. Oocysts were isolated from the feces by discontinuous sucrose gradients followed by microcentrifuge-scale cesium chloride gradients (Arrowood and Sterling, 1987; Arrowood and Donaldson, 1996). Purified oocysts were stored at 4°C in 0.01% Tween 20 solution containing 100 U of penicillin, 100 μg of streptomycin, and 100 μg of gentamicin per mL to retard bacterial growth.

OOCYST ENUMERATION

Stock suspensions of *C. parvum* oocysts were enumerated with a hemocytometer, diluted in deionized water (DI) and ten 100 μL replicate aliquots of the diluted oocyst suspension were placed onto ten individual wells of three welled microscope slides (USEPA, 2001a). The samples were dried (42°C for 1 to 2 h), fixed in methanol, air-dried and a 50-μL aliquot of fluorescein-isothiocyanate conjugated anti-*Cryptosporidium* sp. monoclonal antibody (FITC-CmAb; Waterborne Inc. or Meridian Diagnostics, Inc.) was placed onto each well. The slides were placed in a humid chamber, incubated (37°C for 30 min) and the excess FITC-mAb was aspirated. Any remaining FITC-mAb was removed by adding 50 μL of phosphate buffered saline (PBS; 150 mM, pH 7.2) onto each well, allowing the slides to stand for 1 min and aspirating the excess PBS. This washing step was repeated three times and 50 μL aliquots of a weak DAPI solution (0.4 μg/mL in PBS) were placed onto each well. The slides were allowed to stand at room temperature (2 min) and excess DAPI solution was removed by washing the slides twice in

PBS and once in DI water. The slides were placed in the dark until dry and 10 μL aliquots of mounting medium (2% 1,4- diazabicyclo [2.2.2] octane {DABCO} in 60% glycerol/40% PBS) were placed onto each well, a coverslip applied and the slides examined by using epifluorescence microscopy. Oocysts in seeding trials were less than twelve weeks old (post shedding).

SPIKING METHODS

Spiking and Filtration Procedure for 10 L Sample

A 5 L volume of DI or reagent water was added to a polypropylene carboy with a bottom outlet port and an appropriate volume of the *C. parvum* oocyst working suspensions (containing a known number of organisms) were added. The carboy was placed on a magnetic stir plate and a magnetic stir bar was placed in the carboy and stirred for 2 min before adding another 5 L of DI water. For environmental water matrices, each pellet was diluted in 10 L of DI water and spiked in a similar manner to that described for DI water. Each spiked 10 L sample was mixed for an additional 2 min before proceeding with the filtration procedure. The outlet port of the carboy was connected to a CrypTest cartridge (Whatman, Acton, MA), or an Envirochek standard or HV capsule (Pall Gelman Sciences, Ann Arbor, MI). The outlet of the filtration assembly was plumbed with a flow meter and a Simer pump was used to draw the sample through the filter, with a valve (Alamo Water Treatment No. R5310) regulating the flow to 2 L/min (Figure 2.1).

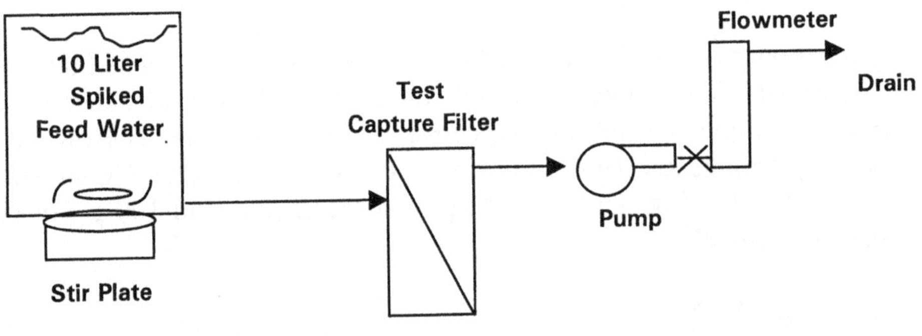

Figure 2.1 Carboy Spiking Arrangement

Spiking Protocols for Fifty Liter Samples

A 500 L bulk sample was collected from the raw water intake of the St. Albans, VT drinking water treatment plant on Lake Champlain. The turbidity was 1.4 NTU. Ten liters were spiked with oocysts and filtered as described above. Following filtration of the spiked 10 L, the carboy was rinsed with 1 L test water and filtration continued. An additional 39 L of the test water was then filtered, for a total sample volume of 50 L. A 50 L volume of unspiked test water was filtered by each of two capsules to serve as matrix blanks for each of the elution procedures.

Spiking Protocols for 1000 Liter Samples

In each experiment, following filtration of 10 L of pre-filtered tap water (flow rate 2 L/min) through each filter, oocysts were spiked directly into the filter. This was followed by filtration of an additional 990 L of pre-filtered tap water through each filter and then elution of each filter according to its respective procedure, as described below. For spike doses greater than 100 organisms, a sub sample (20%) of the eluate from each filter was collected through an individual 25 mm 1 μm pore size PCTE filter, stained with FITC-mAb and analyzed directly. The remaining 80% of eluate from each filter was concentrated with the particle concentrator and the concentrate (15 mL) was subjected to IMS using the Dynal anti-*Cryptosporidium* kit. Where filters were spiked with 50 or 100 oocysts, the entire sample volume was concentrated using the particle concentrator and subjected to Dynal IMS.

Experiments were designed to examine the effect of sodium hexametaphosphate-enhanced elution on the recovery of oocysts spiked into high volume finished water (≤0.1 NTU) samples. Each of the filter capsules was spiked with ~100 oocysts at the beginning of the run, so that a worst-case scenario could be evaluated. Flow rates through each filter were set at 1 L/min, and runs were set up to be 16.6 h, or 1000 min, resulting in 1000 L samples. Flow decay during one series caused total volumes to be 700 L per filter. In addition to the tap water tests, a series of capsules would be spiked and applied to a filtered water matrix to evaluate initial precision and recovery (IPR) of the method. While recoveries from these would be compared to previous data (also spiked at the beginning of sampling), control tests were run which were spiked similarly but underwent the standard elution procedure without the sodium hexametaphosphate step.

SAMPLE ELUTION

Processing of CrypTest Cartridges

The CrypTest cartridges were eluted as follows. Residual water in the housing was poured through the inlet port into a 1 L sample bottle. Tubing was attached to the outlet of the filter housing and the chamber was filled with 80% ICR elution buffer containing 20% of 0.01% Tween-20 solution with the aid of a peristaltic pump (at ~1 L/min). The housing was capped and placed in an ultrasonic bath for 2 min. The eluant was poured through the inlet port into a collection bottle and this backwashing step was repeated one more time and the first and second backwash was combined. The outlet housing was connected to a regulated compressed air source (at 5-10 psi) to displace elution buffer remaining in the housing to the inlet side and this was pooled with the eluate from the two backwashes. The eluate was concentrated by centrifugation at 1050 x g and stored at 4°C for further processing.

Processing of the Envirochek Capsule

After sample filtration, the Envirochek capsule was filled to just above the end of the filter pleats with Laureth-12 buffer. The filter was shaken for 5 min at 80% of maximum speed on a Lab-Line Multi-Wrist-Shaker, with the air vent at the 12:00 or 6:00 position. The eluate was poured off and the capsule was filled again with Laureth-12 buffer and subjected to a further 5 min shaking with the air vent rotated to the 3:00 or 9:00 position. The eluate from the first wash was combined with that of the second wash and analyzed as one sample. Samples were collected in the 250 mL conical centrifugation tubes and centrifuged at 1050 x g for 10 min. The supernatant was aspirated to approximately 1 mL. The pellet was vortexed for 10 sec to resuspend and the pellet volume recorded. The pellet was stored at 4°C for further processing.

Processing of the Envirochek HV Capsule for Source Water

The elution of the HV is similar to that of the original standard Envirochek method, with the exception that following the first 5 min shake and decant, the capsule is rotated 120°, as opposed to 90°. After the second shake, the capsule is not decanted, but rotated an additional 120° to the 240° position. It is then given a third 5 min shake, and the eluate is then decanted to a 250 mL sample tube. This protocol has been adopted for the standard Envirochek in the most recent revision of Methods 1622 and 1623. Each sample was concentrated by centrifugation (1500 x g; 15 min) and the supernatant aspirated to above the pellet. The pellet was vortexed and transferred to a 50 mL conical tube. The 250 mL sample tube was rinsed with phosphate buffered saline with Tween (PBST). The rinse was transferred to the conical tube containing the sample. The sample was diluted to 50 mL with PBST and concentrated by centrifugation (1500 x g; 15 min). The supernatant was aspirated to 10 mL. Pellet material equivalent to 1.0 mL or less was vortexed and transferred to a Leighton tube containing IMS buffers. Where packed pellet volumes exceeded 1 mL in one set of samples, the entire pellet was analyzed by performing IMS on each portion of the pellet. *Cryptosporidium* oocysts were recovered from the interfering debris using IMS and recovered oocysts were stained and enumerated using epifluorescence microscopy according to Method 1622.

Processing of the Envirochek HV Capsule for 1000 L Finished Drinking Water

The method used for sodium hexametaphosphate elution was performed as described in McCuin et al (2001). The capsules were filled with a 5% weight by volume in a DI water solution of sodium hexametaphosphate, and placed on the shaker at the maximum speed for 5 min. The solution was allowed to drain through the filter, so that any materials dissolved or disaggregated could exit through the filter pores, but any oocysts or other intact particles greater than 1 μm would remain in the capsule. This action was followed by a DI rinse, which was poured into the inlet of the capsule and allowed to drain through the filter membrane, continuing the purge of dissolved or sub-micron material. These preliminary steps were followed by the standard HV eluting procedure in its entirety, as described above. The samples were centrifuged and the resulting pellets analyzed by IMS as described in Method 1622.

Processing of the Filta-Max Using the Wash Station

The filter was removed from the filter housing and excess liquid in the housing was poured into a collection beaker. The filter module was screwed onto the base of the plunger located on the IDEXX wash station. The elution tube was screwed in place on the wash station base and the filter module bolt was removed from the filter. The filter was washed with 600 mL of 0.01% PBST by moving the wash station plunger up and down 20 times. The eluate was poured into the collection beaker, and the filter was washed a further 10 times with an additional 600 mL of

PBST. The second eluate was then added to the collection beaker containing the eluate from the first wash.

Sample Concentration Using the IDEXX Particle Concentrator

A 72 mm polysulfone membrane filter with a 3.0 μm pore size was placed in the base of the concentrator, the concentrator tube was screwed in place and the entire unit was placed on a magnetic stir plate. Sample addition to the concentrator tube was followed by attachment of the screw-capped lid (containing a magnetic stir rod) onto the particle concentrator tube. While continuously stirring the sample, a vacuum was applied (2 psi) on the underside of the membrane to concentrate the sample by filtration. The polysulfone filter was then removed, placed in a small stomacher bag containing 5 mL PBST, and stomached for 3 min. The stomacher bag was then agitated on a vortexer for 10 sec, and the resulting eluant was added to a collection bottle. This step was repeated an additional two times, with two 5 mL PBST rinses. All eluates were combined and analyzed as one sample.

Overview of Filta-Max Elution Procedure for Using the Stomacher and Centrifugation

Following the standard procedure for sample collection, the filter module was placed in a large stomacher bag. Using the Allen key provided by the manufacturer, the bolt was removed from the center of the filter allowing the pads to expand in the stomacher bag. The bolt and plastic ends were rinsed with phosphate buffered saline solution with 0.01% PBST and discarded. A volume (600 mL) of PBST was added to the stomacher bag containing the pads. The bag was placed in a large stomacher (Lab Blender, Model 3500) and the pads were washed for 5 min at normal speed. The bag was removed from the stomacher. The wash solution in the bag was decanted into a 2 L pooling beaker. Another 600 mL of fresh PBST was added to the stomacher bag and the washing step was repeated. After the second wash, pads in the bag were removed and retained liquid was expressed from the pads by squeezing them over the pooling beaker. When all pads were removed the remaining solution in the bag was decanted to the pooling beaker. The stomacher bag was rinsed with PBST and these rinses were added to the solution contained in the pooling beaker. The filter wash was transferred to 250 mL conical centrifuge tubes and concentrated at 1500 x g for 15 min. The supernatant was aspirated to approximately 5 mL above the packed pellet. The contents of each 250 mL conical tube was combined in a 50 mL centrifuge tube and further concentrated by centrifugation (1500 x g for 15 min). The supernatant was aspirated to 10 mL. The resuspended pellet was transferred to a Leighton tube for sample clarification.

DYNAL IMS PROCEDURE FOR *CRYPTOSPORIDIUM* OOCYSTS

The IMS procedure was performed as described by the manufacturer (Dynal, A.S., Norway) and approved in Method 1622. Briefly, each 10 mL sample concentrate was added to a Leighton tube containing 1 mL each of 10X SL buffer-A and 10X SL buffer-B. To each tube, 100 μL of the Dynabeads anti-*Cryptosporidium* beads, which were supplied as components of the anti-*Cryptosporidium* IMS kit were added. Samples were incubated for 1 h at room temperature while rotating in an end-over-end rotator. At the end of the incubation period, each Leighton tube was placed in the MPC-1 with the flat side of the tube towards the magnet. The tube was gently rocked for 2 min, through a 90° angle, tilting the tube from the top to the bottom with the magnet side facing up. The supernatant was decanted; each tube was removed from the MPC-1 and 1 mL of 1X SL buffer-A was added to each tube. The tube was gently rocked to re-suspend the bead-oocyst complexes and a Pasteur pipette was used to transfer the suspension into a 1.5 mL Eppendorf tube. Eppendorf tubes were placed in the MPC-M, the magnetic strip was inserted into place and the MPC-M was rocked through 90° for 1 min. The bead-oocyst complex formed a dot on the back of the Eppendorf tube and the supernatant was aspirated. The magnet was removed from the MPC-M and 50 µL 0.1 N HCl was added to each sample and vortexed for a minimum of 10 sec. Samples were allowed to stand for 10 min in the upright position, vortexed for a further 10 sec and the magnet of the MPC-M was inserted into place. The tubes were allowed to stand undisturbed for 2 min. The resulting supernatant was placed in the center of a well of a three-welled microscope slide containing 5 µL 1 N NaOH. A second acid dissociation step was performed. The acidic solution was placed in the same well as the first.

AUREON BIOSYSTEMS IMS PROCEDURE FOR *CRYPTOSPORIDIUM* OOCYSTS

The Aureon *CG* IMS kit (Prod. No.A020302 Aureon BioSystems, GmbH Vienna, Austria) was used to recover *C. parvum* oocysts from finished water matrices as described by the manufacturer. All reagents in the Aureon IMS kit were brought to room temperature before use. Briefly, an amount of each sample concentrate, not exceeding 5 mL and 1 mL packed pellet volume, was transferred to an individual Leighton tube after thoroughly vortexing. If the centrifuge tube was empty, 5 mL A-*IMS* buffer was added to the tube to remove residual debris and transferred to the Leighton tube containing the sample concentrate. If 5 mL of the sample concentrate was a subsample, the 5 mL A-IMS buffer was added directly to the Leighton tube. To each tube, 100 µL of the Aureon *Cryptosporidium* A-*Beads* were added after vortexing. Samples were incubated for 1 h at room temperature while rotating on the sample mixer (25 rpm). At the end of the incubation period, each Leighton tube was placed in the *Magnet*On 4T™ with the flat side of the tube adjacent to the magnet. The tube was gently rocked through a 90° angle, tilting it from top to bottom with the magnet side facing up, for 1 min. The supernatant was decanted; the tube removed from the magnet and 1 mL of A-*Wash* Buffer™was added. The tube was gently rocked to resuspend the bead complex and using a Pasteur pipette, the suspension was transferred to a 2 mL screw-capped cryovial with a rounded bottom. The cryovial was placed in the *Magnet*On 4T and rocked through a 90° motion for 30 sec. The bead-oocyst complex collected as a dot on the back of the vial and the supernatant was aspirated. Another 1mL A-*Wash* Buffer was added to the Leighton to rinse any remaining debris from the tube. Using a Pasteur pipette the rinse buffer was transferred to the same cryovial containing the bead-oocyst complex. The tube was gently rock to resuspend the beads and place back into the *Magnet*On 4T. The magnet was gently rocked through a 90° angle for 30 sec and the supernatant was aspirated. To each tube 50 µL of 0.05 M HCl was added. Each vial was vortexed for 15 sec and allowed to stand undisturbed for 30 sec and vortexed a second time for 15 sec. Each vial was placed in the magnet and allowed to remain undisturbed for 10 sec. The resulting supernatant from each sample was placed in the center of an individual well of a three-welled microscope slide and stained as described below.

STAINING PROCEDURE FOR RECOVERED OOCYSTS

Staining After IMS

After the IMS procedure, the resulting supernatant from each sample was placed in the center of an individual well of a three-welled microscope slide. The samples were dried (42°C for 1-2 h), fixed in methanol, air-dried and 50 μL fluorescein-isothiocyanate conjugated anti-*Cryptosporidium* sp. monoclonal antibodies (Crypt-a-Glo FITC-mAb; Waterborne Inc.) were placed onto each well. The slides were placed in a humid chamber, incubated (37°C for 45 min) and the excess FITC-mAb was aspirated. Any remaining FITC-mAb was removed by adding 50 μL of phosphate buffer saline (PBS; 150 mM, pH 7.2) onto each well, allowing the slides to stand for 1 min and aspirating the excess PBS. Fifty microliter aliquots of a weak 4'6'diamidino-2-phenyl-indole (DAPI) solution (0.4 μg/mL in PBS) were placed onto each well. The slides were allowed to stand at room temperature (2 min) and excess DAPI solution was removed by washing the slides in PBS. The slides were placed in the dark until dry and 10 μL aliquots of mounting medium (2% DABCO in 60% glycerol/ 40% PBS) were placed onto each well, a coverslip applied, sealed with clear nail polish and the slides examined using epifluorescence microscopy.

Direct Detection of Oocysts on 25-mm-Diameter Membranes

A 25-mm-diameter, 1.0-μm-pore-size PCTE membrane filter (Corning, Inc., Acton, MA No. 110610) was housed in a glass chimney filtration assembly (Millipore, Inc., Bedford, MA No. XX10 025 00) and 20% of the filter eluant was passed through by application of a vacuum (15 psi) on the underside of the filter assembly. The inlet side of the glass chimney was rinsed with 5 mL and then 10 mL of 0.01% Tween 20. For each filter an individual humid chamber was constructed by placing a wet paper towel inside a 100 mm x 5 mm plastic Petri dish and cover. A smaller Petri dish (60 mm x 15 mm) cover was inverted on top and a 100-μL aliquot of the *Cryptosporidium* FITC-mAb was placed on top of smaller Petri dish cover. The PCTE membrane filter was placed on top of the FITC-mAb, the large Petri dish cover was applied, and the sample incubated at 35°C for 60 min. Each filter was removed from the humid chamber, placed on the fritted glass support of the filtration assembly and excess FITC-mAb was removed through the filter by application of a gentle vacuum. With the vacuum on, approximately 1 mL PBS was pipetted gently on the filter surface in order to rinse excess FITC-mAb. The filter was placed on a 25 mm x 75 mm slide containing 50 μL of mounting medium. Another 25 μL of mounting medium was placed on top of the filter, a coverslip applied, the filter sealed with clear nail polish and examined by epifluorescence microscopy to quantitatively determine the presence of oocysts.

EPIFLUORESCENCE MICROSCOPY

A Zeiss Axioskop epifluorescence microscope, equipped with a blue filter block (excitation wavelength, 490 nm; emission wavelength, 510 nm) was used to detect FITC-mAb-labeled oocysts at a magnification of 200X. Oocysts were confirmed at 400X magnification by using a UV filter block (excitation wavelength, 400 nm; emission wavelength, 420 nm) for visualization of DAPI. Internal morphology of oocysts was observed by using Nomarski differential interference contrast (DIC) optics microscopy.

PBMS TIER 1 AND TIER 2 VALIDATION STUDIES

Laboratories

Five laboratories with expertise in *Cryptosporidium* analysis using Method 1622 were used for the collaborative testing required under PBMS. The laboratories were CH Diagnostic and Consulting Service, Inc. (Loveland, CO), UL-Environmental Health Laboratories, Inc. (South Bend, IN), the New York City Department of Environmental Protection (NY), and the Metropolitan Water District of Southern California (LaVerne, CA), and Clancy Environmental Consultants, Inc. (CEC).

Samples

Each laboratory used its own laboratory reagent water and source water as samples.

Preparation of Spike Doses for *Cryptosporidium*

Spike dose suspensions were obtained from the Wisconsin State Laboratory of Hygiene. These suspensions were prepared by flow cytometry according to protocols used by USEPA in their performance evaluation studies and required by USEPA for method validation studies. Suspensions were sent to the laboratories by overnight courier. Each laboratory received 10 mL centrifuge tubes containing an unknown number of oocysts. A single vial was used for each test filter.

Experimental Format for Validation Study

For Tier 2 validation, a total of eight samples were spiked and processed in each laboratory (five initial precision and recovery samples and three matrix spike samples). For Tier 1 validation, the five IPR samples were prepared an analyzed, and in some cases, a set of MS/MSD samples was also analyzed. The content of a single centrifuge tube was added to a 50 L sample contained in a carboy using the protocol described previously. The seeded samples were filtered as described above for each filter.

After sample seeding and analysis, raw data sheets were sent to CEC for data analysis and interpretation using the PBMS format. Formal reports were sent to the USEPA for studies that met the method acceptance criteria.

Figure 2.2 is a flow diagram showing the various method components that were evaluated in this study and the PBMS validation outcomes.

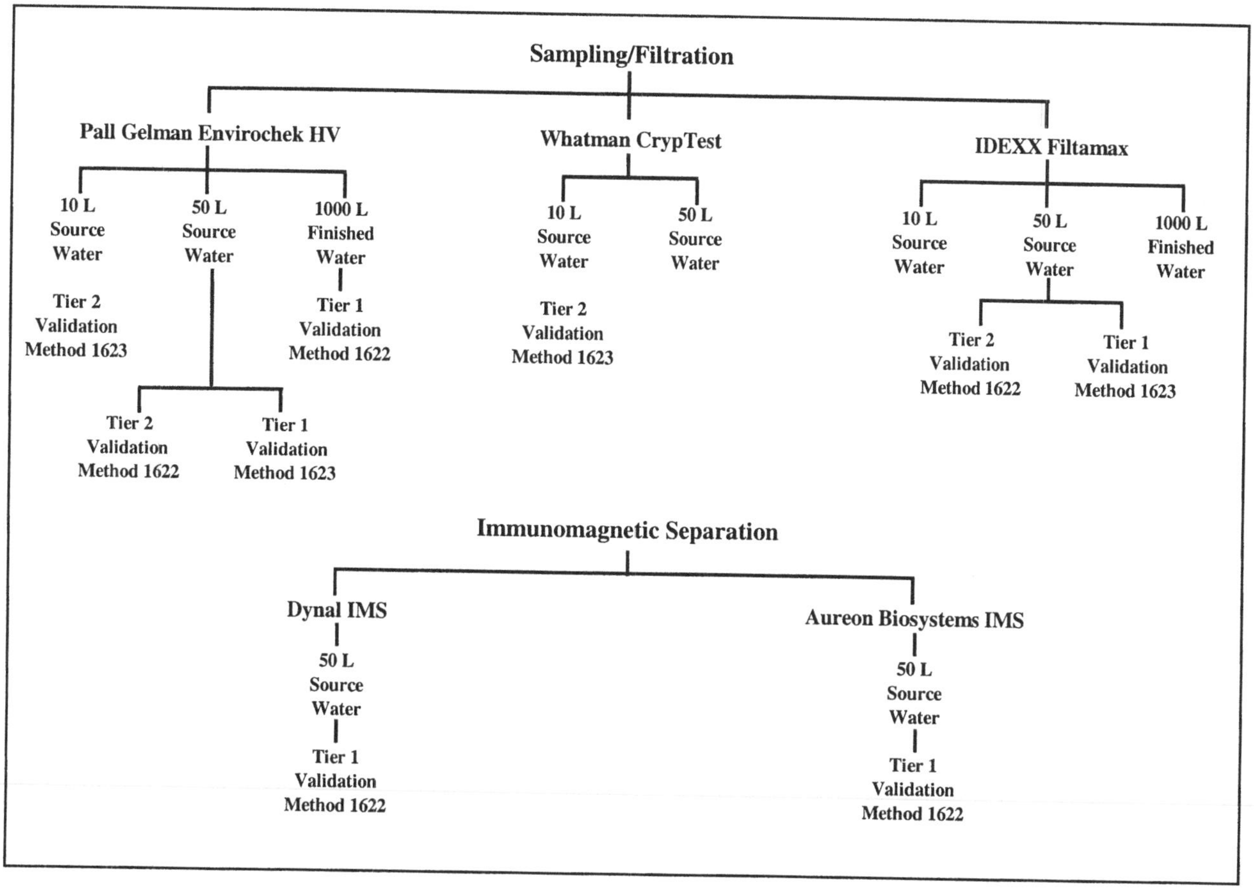

Figure 2.2 Flow Diagram Showing the Various Method Components Evaluated and Validated

CHAPTER 3
RESULTS AND DISCUSSION

SOURCE WATER SAMPLING

The project sought to develop high volume methods for both source and finished drinking water samples. Methods 1622 and 1623 were limited to 10 L sample volumes, and in the U.S. there was no method other than the ICR method for high volume finished drinking water analysis. In this project, two high volume methods for 50 L source water samples were developed and validated using PBMS. Two high volume finished water methods were developed and single laboratory-validated for 1000 L samples.

High Volume (50 L) Source Water Methods

The Envirochek HV Capsule

The Pall Gelman Envirochek capsule was adopted by the USEPA in Methods 1622 and 1623 for use with 10 L sample volumes. When larger volumes were sampled with the Envirochek, recoveries were noted to be lower. To address the issue of high sample volume, Pall Gelman developed a second product, the Envirochek HV capsule. The HV is identical to the standard Envirochek except that the pleated filter in the capsule is a polyester track-etch membrane material in place of polysulfone. Studies were undertaken to develop performance data for use of the HV for 50 L source water samples, and once this was complete, the HV was evaluated for Tier 2 validation for use in Method 1622 in place of the standard Envirochek for collecting samples volumes up to 50 L.

A collaborative trial using the PBMS criteria developed by the USEPA (USEPA, 1996) was designed and conducted as described in the EPA Guide to Method Flexibility and Approval of EPA Water Methods (USEPA, 1997). These criteria are used in PBMS studies to determine method equivalency from the data developed by USEPA in the Supplemental Surveys (USEPA, 2001e). Three laboratories received blind spike vials of live *Cryptosporidium* oocysts prepared by flow cytometry. Each laboratory prepared and analyzed eight samples. Five samples were reagent water - four seeded with *Cryptosporidium* oocysts and a fifth seeded with a blank; three samples were 50 L source water, two seeded with *Cryptosporidium* oocysts and the third seeded with a blank vial containing no oocysts. Laboratories did not know which vials contained oocysts nor at what level until the results were reported to the CEC study manager. The protocol used to spike the samples was that used in the USEPA 1622 validation studies and described in Method 1622. Contents of the vials were added directly to the 50 L volume as it was being stirred, and stirring continued throughout the filtration.

Samples were eluted and processed as described in Materials and Methods. Pellet material equivalent to 1 mL or less was vortexed and transferred to a Leighton tube containing IMS buffers. Where packed pellet volumes exceeded 1 mL, the entire pellet was analyzed in 1 mL aliquots. *Cryptosporidium* oocysts were recovered from the interfering debris using IMS and recovered oocysts were stained and enumerated using epifluorescence microscopy according to Method 1622. The validation data are presented in Table 3.1. The Envirochek HV with 50 L source water samples achieved equivalent performance to the standard Envirochek with 10 L (USEPA, 2001c, d). The data from the three participating laboratories fell within the acceptance criteria for the method as shown in Table 1.1.

Table 3.1
IPR and MS/MSD Validation Data for *Cryptosporidium* Recovery Using Method 1622 and the Envirochek HV with 50 L Source Water Samples

Sample Description	Lab	Sample Turbidity (NTU)	Packed Pellet Volume (mL)	Spike Dose (#)	No. Oocysts Recovered	Background Adjusted Count	Percent Recovery	Mean % Recovery	RSD/ RPD (%)
Reagent Blank	1	<0.1	<0.1	0	0	NA			
IPR1	1	<0.1	<0.1	96	43	43	44.9		
IPR2	1	<0.1	<0.1	96	48	48	50.1		
IPR3	1	<0.1	<0.1	96	43	43	44.9		
IPR4	1	<0.1	<0.1	96	38	38	39.6	44.9	9.5
Reagent Blank	2	0.05	<0.05	0	0	NA			
IPR1	2	0.05	<0.05	96	78	78	81.4		
IPR2	2	0.05	<0.05	96	69	69	72.0		
IPR3	2	0.05	<0.05	96	53	53	55.3		
IPR4	2	0.05	<0.05	96	59	59	61.6	67.6	17.0
Reagent Blank	3	0.03	0.01	0	0	NA			
IPR 1	3	0.03	0.01	96	45	45	47.0		
IPR 2	3	0.03	0.01	96	53	53	55.3		
IPR 3	3	0.03	0.01	96	68	68	70.9		
IPR 4	3	0.03	0.01	96	66	66	68.9	60.5	18.8
IPR Acceptable Range								**24-100**	**<55**
Matrix Blank	1	10.6	3.0	0	0	NA			
MS1	1	9.4	3.0	96	28	28	29.2		
MS2	1	10.1	4.0	96	28	28	29.2	29.2	0.
Matrix Blank	2	2.1	0.5	0	0	NA			
MS1	2	2.1	0.5	96	62	62	64.7		
MS2	2	2.1	0.5	96	76	76	79.3	72.0	20.3
Matrix Blank	3	1.8	0.45	0	0	NA			
MS 1	3	1.8	0.45	96	57	57	59.5		
MS 2	3	1.8	0.45	96	51	51	53.2	56.3	11.1
Matrix Acceptable Range								**13-111**	**<61**

IPR = Initial Precision and Recovery
MS/MSD = Matrix Spike and Matrix Spike Duplicate
RSD = Relative Standard Deviation/RPD = Relative Percent Difference

The Filta-Max Filter

The IDEXX Filta-Max filter was validated for 50 L source water samples using the protocol described previously for spiking. The samples were processed according to Method 1622 using the filter elution protocol described in Chapter 2. Validation data are shown in Table 3.2. The data from the three participating laboratories fell within the acceptance criteria for the method as shown in Table 1.1.

Table 3.2
IPR and MS/MSD Validation Data for *Cryptosporidium* Recovery Using Method 1622 and the Filta-Max with 50 L Source Water Samples

Sample Description	Lab	Sample Turbidity (NTU)	Packed Pellet Volume (mL)	Spike Dose (#)	No. Oocysts Recovered	Background Adjusted Count	Percent Recovery	Mean % Recovery	RSD/ RPD (%)
Reagent Blank	1	<0.1	0.01	0	0	NA	0		
IPR1	1	<0.1	0.01	101	51	51	0		
IPR2	1	<0.1	0.01	101	49	49	51		
IPR3	1	<0.1	0.01	101	45	45	49		
IPR4	1	<0.1	0.01	101	58	58	45	50.3	10.7
Reagent Blank	2	<0.1	0.01	0	0	NA			
IPR1	2	<0.1	0.01	103	64	64	62.4		
IPR2	2	<0.1	0.01	103	51	51	49.7		
IPR3	2	<0.1	0.01	103	43	43	41.9		
IPR4	2	<0.1	0.01	103	47	47	45.8	50.0	17.8
Reagent Blank	3	<0.1	0.01	0	0	NA			
IPR 1	3	<0.1	0.01	103	55	55	53.6		
IPR 2	3	<0.1	0.01	103	51	51	49.7		
IPR 3	3	<0.1	0.01	103	55	55	53.6		
IPR 4	3	<0.1	0.01	103	50	50	48.7	51.4	5.0
IPR Acceptable Range								**24-100**	**<55**
Matrix Blank	1	1.0	0.1	0	0	NA			
MS1	1	1.0	0.1	101	55	55	54.5		
MS2	1	1.0	0.1	101	43	43	42.6	48.6	24.5
Matrix Blank	2	4.8	0.4	0	0	NA			
MS1	2	4.8	0.4	103	24	24	23.4		
MS2	2	4.8	0.4	103	20	20	19.5	21.4	18.2
Matrix Blank	3	7.1	0.5	0	3	NA			
MS 1	3	6.3	0.4	103	53	50	48.7		
MS 2	3	7.2	0.5	103	54	51	49.7	49.2	2.0
Matrix Acceptable Range								**13-111**	**<61**

IPR = Initial Precision and Recovery
MS/MSD = Matrix Spike and Matrix Spike Duplicate
RSD = Relative Standard Deviation/RPD = Relative Percent Difference

Both the Pall Gelman Envirochek HV and the IDEXX Filta-Max are now included in Method 1622 as validated components using the PBMS criteria for method improvement. These filters were validated with 50 L samples in place of the standard 10L sample size, permitting better method sensitivity for those utilities that conduct routine sampling for *Cryptosporidium*.

Tier 1 Validation: Evaluation of the Stomacher for Elution of the Filta-Max

The objective of this study was to assess overall method recoveries using the stomacher method for eluting the Filta-Max using criteria specified in PBMS for Tier 1 validation (Tables 1.1 and 1.3). Tier 1 PBMS validation involves a single laboratory demonstrating that a modification to a validated method produces equivalent results. If equivalent results are obtained the modification can be incorporated into the method for all in-house work. The study was undertaken to develop an alternate method of washing the foam filters that would be faster and use disposables to minimize contamination or carryover from sample to sample. The stomacher was used for sample elution, followed by centrifugation for sample concentration; this technique replaces the standard elution method using the wash station and particle concentrator. The IDEXX wash station is difficult to clean with assurance that no oocysts are carried over to subsequent samples; the use of the stomacher is faster and uses disposable bags for sample elution.

Three laboratories conducted Tier 1 validation studies. Raw IPR and MS/MSD results are presented in Table 3.3. Mean oocyst recovery in reagent water was 49.4% with recoveries ranging from 40.9% to 62.1%. All data fell within the method acceptance criteria, indicating that this modification is robust.

Table 3.3
Summary of Independent Tier 1 Method 1622 Modification Validation Data Replacing the Standard Elution Method with Stomaching the Foam Pads and Centrifugation of the Eluate

Sample Type	Lab	Turbidity (NTU)	Spike Dose	Percent Recovery	Mean Percent Recovery	RSD/RPD
IPR1	1		95	45.3		
IPR2	1		95	40.0		
IPR3	1		95	41.0		
IPR4	1		95	44.2	42.6	5.9
MS1	1	4.6	95	30.5		
MS2	1	4.3	95	32.6	31.6	6.7
IPR1	2		101	68.1		
IPR2	2		101	55.3		
IPR3	2		101	70.1		
IPR4	2		101	51.3	61.2	15.2
MS1	2	2.5	101	36.5		
MS2	2	2.6	101	35.5	36.0	2.7
IPR1	3		146	47.4		
IPR2	3		146	37.8		
IPR3	3		146	48.7		
IPR4	3		146	35.7	42.4	15.6
MS1	3	4.9	146	17.2		
MS2	3	4.9	146	16.5	16.8	4.1
IPR Acceptance Criteria (Method 1623, April 2001)					**24-100%**	**55%**
Matrix Acceptance Criteria (Method 1623, April 2001)					**13-111%**	**61%**

IPR = initial precision and recovery
MS = matrix spike
RSD/RPD = relative standard deviation/relative percent difference

The method modification using the stomacher/centrifugation technique is an acceptable alternative to the wash station for elution of *C. parvum* oocysts from the Filta-Max filter. This method is far less cumbersome to use, both in ease of operation and timesavings. One concern of Filta-Max users is the need to use chlorine to inactivate the surface epitopes of oocysts and cysts to prevent cross contamination of samples. Chlorine is a very corrosive agent that can weaken the plastic parts of the wash station and in turn, can potentially cause leakage and loss of sample. The wash station also requires increased hands-on time to elute the filter. The stomacher technique is an alternative that offers reduced hands-on time to elute the filter and minimizes clean up between samples. Cross contamination issues with the wash station are eliminated with the use of the disposable stomacher bag.

Table 3.4 shows the CEC single laboratory ongoing performance data with Filta-Max using the Stomacher modification, showing that *Cryptosporidium* recoveries are high and reproducible over time. OPR is a requirement of Methods 1622 and 1623, and these are the spiked reagent water samples run weekly in the lab to establish that the method is performing within its specifications.

Table 3.4
Single Laboratory IPR and OPR Data for Oocyst Recoveries in 10 L Reagent Water Samples Using the Filta-Max and USEPA Method 1622

Sample ID	Oocyst Spike Dose (Mean ±SD)	Number of Oocysts Recovered	Oocyst Recovery (%)	Percent Oocyst Recovery (Mean ± SD)
IPR 1	170± 3	84	49.5	
IPR 2		62	36.6	41.9 ± 6.0
IPR 3		74	43.6	
IPR 4		64	37.7	
OPR 1	110 ± 1	48	43.5	
OPR 2		52	47.1	47.9 ± 3.5
OPR 3		51	46.2	
OPR 4		58	52.5	
OPR 5	99 ± 1	48	50.1	

No oocysts were detected in method blanks.
IPR = initial precision and recovery
OPR = ongoing precision and recovery

FINISHED WATER SAMPLING

Evaluation of Filters for *C. parvum* Oocyst Recoveries in High Volume Pre-filtered Tap Water Samples

The Envirochek HV, CrypTest and Filta-Max filters are all approved for use for sample collection in USEPA Method 1622. The Filta-Max and Envirochek HV filters are approved by the U.K. Drinking Water Inspectorate for use in their continuous monitoring program for determining presence of *Cryptosporidium* oocysts in large volumes of finished water. As these filters offered the potential to be adapted for collection of high volumes of finished water samples, they were the obvious choice for initial assessment of their suitability for collecting large volume samples. Preliminary work was conducted on filtered tap water (filtered serially through two 1-µm yarn wound filters, and a 1 µm precise pore size PCTE filter), which served to reduce the levels of contaminating debris. The objective was to ascertain the impact of filtering large sample volumes on oocyst recoveries with the existing elution procedures. Each filter was evaluated at various spike doses, ranging from 50 to 1000 oocysts. Enumeration data of each spike dose has been summarized in Table 3.5.

Table 3.5
Spike Doses of *C. parvum* Used for Assessing Recoveries from Envirochek HV, CrypTest and Filta-Max Filters and 1000 L Volumes of Filtered Tap Water

Spike Dose	Trial 1	Trial 2	Trial 3
50	51.6 ± 5.5	58.1 ± 11.4	52.1 ± 6.7
100	91.9 ± 9.0	101.1 ± 9.3	92.3 ± 9.5
200-300	315.8 ± 18.0	311.7 ± 26.3	
500	407.6 ± 24.4	438.4 ± 44.4	488.2 ± 28.3
1000	980.0 ± 42.0	909.2 ± 70.8	

Figures 3.1 and 3.2 indicate *Cryptosporidium* oocyst recovery efficiencies by both direct analysis and the overall method involving IMS separation. Figure 3.1 shows that the direct method of analysis (avoiding IMS) achieved higher recoveries than when 80% of the filter eluate was subjected to overall analysis. Recoveries with the three filter types ranged from 21% to approximately 51%. Although the largest spike dose (1000 oocysts) indicated generally low oocyst recoveries, spike doses of 500, 300, 100 and 50 oocysts did not reveal marked difference in oocyst recoveries. Using the selected spike doses, the initial oocyst concentration in the samples did not appear to impact overall recoveries (Figure 3.2).

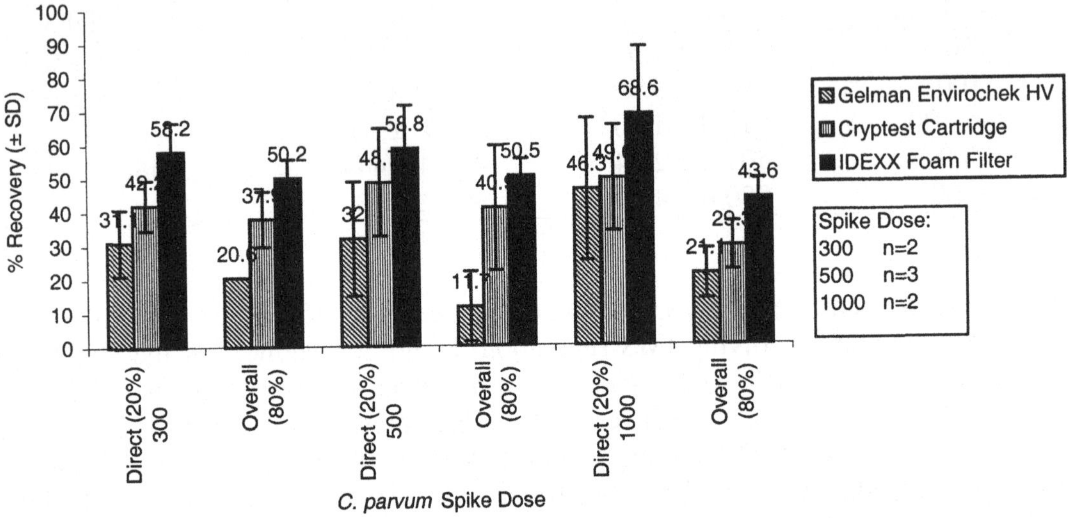

Figure 3.1. ***C. parvum*** **Oocyst Recoveries from 1000 L of Pre-Filtered Tap Water Comparison of Various Filters and Direct versus Overall Procedures.**

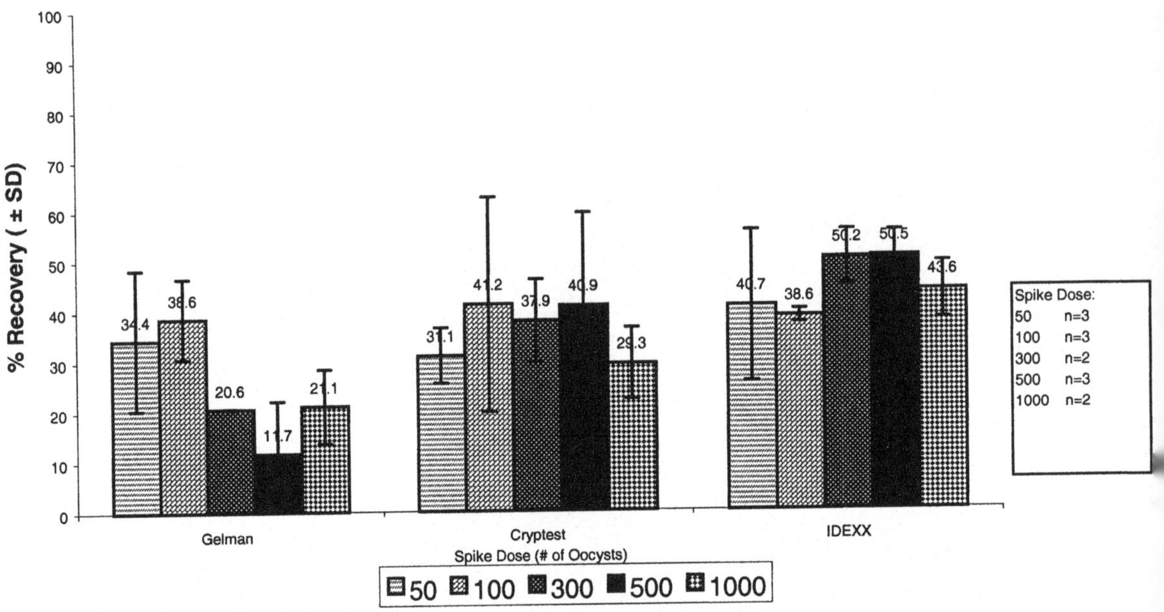

Figure 3.2. Effect of Spike Doses on ***C. parvum*** **Oocyst Recoveries from 1000 L of Pre-filtered Tap Water.**

Evaluation of Filters for Oocyst Recovery in 1000 L Tap Water Samples

Although the initial experiments were conducted to determine the impact of collecting large sample volumes on oocyst recoveries with the three filter types using pre-filtered tap water, it was imperative to assess performance of these filters using large volumes of tap water. Preliminary trials of tap water supplied in St. Albans, VT were conducted using the Envirochek HV, CrypTest, and Filta-Max filters. Samples were collected in the laboratory at the sink tap, and represent delivered distribution system water. Each filter was tested in duplicate at a spike dose of 500 oocysts. In each experiment, 10 L of tap water was initially sampled (2 L/min) through the filter. At this point, the filters were spiked (directly into the filter) with a known amount of oocysts (375.7 ± 43.7 oocysts and 590.2 ± 68.0 oocysts for replicates 1 and 2, respectively). An additional 990 L of water was then passed through the filter (2 L/min). The samples were eluted and concentrated according to their respective procedure, as described in Chapter 2. The resulting samples were subjected to Dynal IMS, labeled with anti-*Cryptosporidium* FITC-mAb, and examined by epifluorescence microscopy. The overall oocyst recoveries are presented in Table 3.6 and were less than 5% with all three filters.

Table 3.6
Overall Recoveries of *C. parvum* Oocysts from 1000 L Tap Water Samples

Filter Type	*C. parvum* Oocysts % Recovery ± SD
Gelman Envirochek HV	3.9 ± 2.4
Whatman CrypTest	3.5 ± 1.1
IDEXX Filta-Max	4.8 ± 1.6

Cryptosporidium oocyst recoveries with the Envirochek HV, CrypTest or Filta-Max filters were not impacted by collection of large sample volumes of pre-filtered finished water. However, when the pre-filtered water samples were replaced with tap water samples, oocyst recoveries demonstrated a dramatic decline with all three filters. This was intriguing, as the Filta-Max has been used in the U.K. for their continuous monitoring program for large volumes of finished water for over two years. Data from the U.K. (personal communication) indicate that these filters are capable of yielding good over-all recoveries (i.e. 30-50%), which was also supported by our investigation using pre-filtered tap water. This led us to speculate that recoveries in our trials may have been impacted by either differences in water quality, perhaps as a result of differences in the chemical composition of the finished water. For example, presence of varying concentrations of flocculants in the finished water could impact overall recoveries. Where these flocs are present in high concentration in tap water, their collection on the test filters may provide greater adhesion of organisms (either physically or chemically) to the matrix in the sample collection filters. Another possibility was the presence material from the distribution system in samples collected at the lab tap. Should this be the case, it would necessitate rigorous elution protocols to recover the organism than what has been employed presently. With this rationale, if the organism entrapment is physical then employing more rigorous elution protocols could help to improve oocyst recoveries; however, it was also likely that the organisms were being retained on the filter matrix by chemical entrapment. If this were the case, then providing an environment for reversal of this process would lead to greater improvements in organism recovery.

Effects of Tap Water Concentrates on Performance of Dynal Dynabeads anti-*Cryptosporidium* IMS

While recoveries of the Dynabeads anti-*Cryptosporidium* IMS beads have been well characterized with respect to performance in DI water and source water concentrates (Bukhari et al, 1998; Rochelle et al, 1999), it was unclear how performance of these beads might be affected in concentrates generated from tap water. In order to rule out that it was the filtration/elution step and not the IMS step that was responsible for the low recoveries noted in the high volume data in Table 3.6, it was necessary to examine how the IMS performed with finished drinking water concentrates. To address this, finished water concentrates were generated by collection of large volumes of St. Albans tap water (2,500 L) through the Filta-Max filters. The filters were eluted, the eluate concentrated to 50 mL, and divided into five 10 mL sub-samples. Each 10 mL concentrate represented 500 L of finished water and contained a packed pellet volume of 0.4 mL. Five 10 mL finished water concentrates and duplicate 10 mL DI water control samples were each spiked with 98.1 ±8.6 oocysts and subjected to the Dynabeads IMS as described by the manufacturer.

Data from the tap water concentrates shown in Table 3.7 indicated recoveries ranging from between 66 to 79%. Although the mean oocyst recoveries in tap water were lower than in DI water, it is likely that these differences were due to limited recovery trials in the DI water control samples in this study. However, the performance of Dynabeads anti-*Cryptosporidium* in DI water has been well characterized previously, and the oocyst recovery range obtained in these finished water trials was within that expected range. This suggests that the chemical composition of St. Albans tap water and physical characteristics of the matrix generated from 500 L volumes did not compromise the IMS performance significantly.

Table 3.7
***C. parvum* Recoveries from Tap Water Concentrates Using Dynal Dynabeads anti-*Cryptosporidium* IMS**

Replicate No.	% Recovery	
	DI Water	Tap Water Concentrates
1	77.5	75.4
2	92.8	79.5
3		69.3
4		72.4
5		66.3
Mean ± SD	85.2 ± 10.8	72.6 ± 5.1

Mean *C. parvum* oocyst spike: 98 ± 9

High Volume (1000 L) Finished Drinking Water Methods

Two high volume finished water methods were developed and single laboratory-validated for 1000 L samples. The method employing the Filta-Max was the method developed for the U.K. Drinking Water Inspectorate continuous monitoring regulation. The method employing the Envirochek HV required a new approach in eluting the oocysts from the filter itself. Finished water samples were those collected at the treatment plant, while tap water samples were collected at the laboratory tap, and hence not subject to distribution system effects.

The Envirochek HV Capsule

It was recognized that when using the standard elution procedures for the Envirochek HV capsules following filtration of a high volume (500-1000 L) of finished water, residual material remained on the filter surfaces, and that extraordinary means would be necessary to complete the elution. Various steps were taken to improve elution, such as increasing the duration of the agitation steps for the three filter types, using more aggressive washing procedures, and by acidifying the elution solution in the hope of dissolving or breaking up the layer in which oocysts were trapped. None of these measures produced improved recoveries.

Sodium hexametaphosphate has been known to reduce scaling for some time (Fireman and Reitemeier, 1944). It is the main ingredient used in household products to aid cleansing in hard water conditions. The chemical has even found application in reducing plaque formation on dog's teeth (Stookey et al, 1995). In drinking water applications, it is used as a scale inhibitor in membrane pre-treatment (Schock, 1990; Kasper, 1993).

The use of sodium hexametaphosphate to dislodge, dissolve, disaggregate, or otherwise release oocysts from the capsule membrane surface was successful. Results of earlier tests of the Envirochek HV capsule are shown in Table 3.8. From a filtered tap water matrix, average recoveries of 36% were achieved, but with a fairly wide range (16-61%). Tests in tap water without sodium hexametaphosphate resulted in only 2-6% recovery. In this study (Table 3.9), including the sodium hexametaphosphate step, an average of 62% was recovered from filtered water, with a narrower range (58-68%). The results from an actual finished water matrix show dramatic enhancement with the sodium hexametaphosphate, boosting recoveries of oocysts to 38% from one series using 700 L samples, and 45% from 1000 L samples. Spiked controls (Table 3.9) run without sodium hexametaphosphate in the present study confirmed the previously noted poor recovery, allowing only 4-7%.

Table 3.8
Recoveries of *Cryptosporidium* Oocysts Using the Standard Method 1622 Gelman Envirochek HV Capsule Elution Method

Sample Volume (L)	Matrix	Number of Tests	Average Percent Recovery	Range Percent Recovery
1000	Filtered tap water	13	36	16-61
1000	Tap water	2	4	2-6

Table 3.9
Recoveries of *Cryptosporidium* Oocysts Using the Sodium Hexametaphosphate Envirochek HV Capsule Elution Method

Sample Volume (L)	Matrix	Number of Tests	Average Percent Recovery	Range Percent Recovery
1000	Filtered tap (IPR)	4	62.2	58-68
700	Tap water	3	37.7	35-41
1000	Tap water	3	45.6	37-53
1000	Tap water (standard elution protocol, no sodium hexametaphosphate)	2	6.0	4.5-7.4

IPR = Initial precision and recovery

During the elution steps, the tap water samples that had been pre-treated with sodium hexametaphosphate were visibly altered from the non-treated samples. The sodium hexametaphosphate solution that was drained out of the capsule was very discolored by dissolved material. The resulting pellet volume of the untreated samples were approximately 200 μl, but contained only a trace (<10 μl) when pre-treated. These observations indicate that significant material was dissolved or at least broken up into small particles and eliminated from the eluate, freeing the oocysts for subsequent separation procedures.

Tier 1 Validation of the Envirochek HV for 1000 L Finished Drinking Water Samples

CEC conducted a Tier 1 validation of the high volume finished drinking water method using the Envirochek HV capsule. The data are presented in Table 3.10 and show that the high volume finished drinking water method meets the EPA acceptance criteria for *Cryptosporidium* recovery, and can be used reliably in these analyses.

Table 3.10
Tier 1 Validation Data for *Cryptosporidium* Recovery Using Method 1622 and the Envirochek HV for 1000 L Finished Drinking Water Samples with Sodium Hexametaphosphate Elution

Sample Description	Sample Turbidity (NTU)	Spike Dose (#)	No. Oocysts Recovered	Background Adjusted Count	Percent Recovery	Mean % Recovery	RSD or RPD
Reagent Blank	<0.1	0	0	NA			
IPR1	<0.1	99	68	68	68.5		
IPR2	<0.1	99	58	58	58.4		
IPR3	<0.1	99	60	60	60.4		
IPR4	<0.1	99	61	61	61.4	62.2	7.0
IPR Acceptable Range in Source Water						**13-143**	**<67**

IPR = Initial Precision and Recovery
RSD = Relative Standard Deviation/RPD = Relative Percent Difference
NA = Not Applicable

It was uncertain whether the advantages of sodium hexametaphosphate for elution of finished water matrices might be realized with source water samples. These samples presently are low volume (10 to 50 L) and sampled over relatively short time periods, and are less apt to form a solidifying cake on the membrane surface. Under the ICR Supplemental Survey using Method 1623, matrix spike recoveries in 10 L indicated the polyethersulfone capsule filter gave average *Cryptosporidium* recoveries of 43% (Connell et al, 2000) in utility source water. In another study, however, using Method 1622, the Envirochek capsule recovered only 12% of *Cryptosporidium* oocysts spiked in surface (stream) waters (Simmons et al, 2001). The research team decided to look at sodium hexametaphosphate to see if oocyst recoveries could be improved in high volume (50 L) source water samples.

Six Envirochek HV capsules were seeded with *Cryptosporidium* oocysts spiked into 50 L samples of source water from Lake Champlain with a turbidity of 1.4 NTU. Following filtration, three of these capsules were eluted using the standard HV procedure. The other three capsules were eluted by incorporating enhancement steps for the sodium hexametaphosphate adaptation to Method 1622, and the resulting recoveries compared.

Results are given in Table 3.11. The recoveries using the standard HV elution and the enhanced steps were not significantly different. The pronounced gain in recovery due to the sodium hexametaphosphate steps that had been realized for 1000 L finished water samples (from less than 10% to greater than 40%) was not evident with the source water samples. It is postulated that the diminished recovery noted in finished water may be due to the presence of chemicals or particles not present in the source water, so that no advantage is gained with the sodium hexametaphosphate, if it is acting to inhibit scaling.

Table 3.11
Comparison of Two Eluting Methods for Envirochek HV Capsules with 50 L Source Water Samples

Sample Volume (L)	Matrix/Elution Solution	Number of Tests	Average percent Recovery	Percent Recovery Range
50	Source water/standard Laureth-12 buffer	3	52 ±6	45.6 – 56.1
50	Source water/ sodium hexametaphosphate buffer	3	48 ±8	39.2 – 54.0

No oocysts were recovered from the matrix blank samples.

Filta-Max for Finished Drinking Water

The Filta-Max is used routinely in the UK for continuous monitoring of finished drinking water as required by the UK Drinking Water Inspectorate. Thousands of samples are analyzed each year using the Filta-Max for collection of 1000 L finished water samples. The research team developed additional data on the use of the Filta-Max showing that it performs very well in oocyst capture and recovery from 1000 L samples using the Method 1622 protocol. Table 3.12 shows that recoveries remain high and consistent using spike doses as low as 50 oocysts. The data meet the Method 1622 acceptance criteria (mean recovery ranging from 24 to 100%, standard deviation at ≤ 55%).

Table 3.12
Recovery of *Cryptosporidium* Oocysts in 1000 L Finished Water Samples Using the Filta-Max System

Approximate Concentration of *Cryptosporidium* Oocysts	Percent Recovery (Mean ± SD)
50 (n=3)	40.7 ± 21.1
100 (n=3)	38.6 ±17.3
300 (n=2)	50.2 ± 5.5
500 (n=3)	50.5 ± 22.9
1000 (n=2)	43.6 ±5.7

IMMUNOMAGNETIC SEPARATION

Several IMS kits have been developed in the last decade for separation of *Cryptosporidium* oocysts from water samples, but to date only two are commercially available and in routine use. These are the Dynal Dynabeads anti-*Cryptosporidium* and anti-*Giardia* kits and the newly introduced Aureon Biosystems IMS kits for *Giardia* and *Cryptosporidium* in water. For the majority of the work done in this study the Dynal kit was used, but once the Aureon kit became commercially available, studies were done to assess its efficacy.

Tier 1 Validation of the Aureon IMS Kit

The Aureon kit was tested by CEC and two other laboratories and found to be equivalent to the Dynal kit (Tier 1 validation) for recovery of *Cryptosporidium* in Method 1622. Tier 1 does not require matrix spikes but these were done in some instances. Data are presented in Table 3.13.

Table 3.13
Tier 1 Validation Results of the Aureon IMS Kit in Methods 1622 with 50 L Source Water Samples

Sample ID	Lab	Spike Dose	Turbidity (NTU)	Oocyst Count	Percent Recovery	Mean Percent Recovery	RSD/ RPD
IPR 1	1	150		57	38.0		
IPR 2	1	150		60	40.0		
IPR 3	1	150		48	32.0		
IPR 4	1	150		76	50.6	40.1	19.4
MS 1	1	150	12.0	32	21.3		
MS 2	1	150	10.0	30	20.0	20.6	6.5
MS 3	1	0	11.0	0			
IPR 1	2	150		97	64.8		
IPR 2	2	150		94	62.8		
IPR 3	2	150		93	62.1		
IPR 4	2	150		87	58.1	61.9	4.5
MS 1	2	150	2.2	106	70.8		
MS 3	2	150	2.2	85	56.7	63.8	22.0
MS 2	2	0	2.2	0			
IPR 1	3	150		27	18.0		
IPR 2	3	150		46	30.7		
IPR 3	3	150		72	48.1		
IPR 4	3	150		77	51.4	37.1	42.9
IPR Acceptance Criteria for Crypto. (Method 1623, April, 2001)						**24-100%**	**49%**
Matrix Acceptance Criteria for *Crypto.* (Method 1623, April 2001)						**13-111%**	**61%**

IPR = initial precision and recovery
MS = matrix spike
RSD/RPD = relative standard deviation/relative percent difference

CEC also compared the Aureon kit to the Dynal IMS kit for recovery of oocysts in 1000 L finished drinking water samples and found it to be equivalent as shown in Table 3.14.

Table 3.14
Comparison of Dynal and Aureon IMS Kits for *Cryptosporidium* Recovery in 1000 L Finished Drinking Water Samples

IMS Kit	Oocyst Spike Dose	Percent Recovery	Mean Percent Recovery	RSD
Aureon	107 ± 9	77.4		
		58.8		
		78.4	71.5	11.1
Dynal	107 ± 9	70.9		
		66.2		
		63.4	66.9	3.8

The testing conducted by CEC and the participating labs indicate that the Aureon IMS kit is equivalent to the Dynal IMS kit using the EPA's Tier 1 PBMS criteria. Both kits employ the Waterborne mAbs for *Cryptosporidium*; only the bead conjugation is different.

CHAPTER 4
SUMMARY AND CONCLUSIONS

The objectives of this study were to develop methods for high volume sampling of source and treated water, and to collaboratively test the new methods so that they can become USEPA standards for *Cryptosporidium* analysis of water. These objectives were met successfully.

The methods developed for high volume analysis of source and treated water were adaptations of USEPA Method 1622, which was originally developed for 10 L source water analysis. Method 1622 was improved in this study to accommodate 50 L sample volumes. Two filters, the IDEXX FiltaMax system and the Pall Gelman Envirochek HV, were both validated for 50 L source water samples using the USEPA PBMS criteria. Permitting larger samples to be collected allows more sample volume to be analyzed and increases method sensitivity. The performance criteria for 10 L samples was met using 50 L sample volumes, thereby increasing sensitivity with no loss in method performance.

Another important outcome of this research was the development of a high volume finished drinking water method, capable of handling 1000 L samples. Prior to this, the only method for use in the U.S. for finished drinking water testing was the ICR method, which is recognized to be poor (Clancy et al, 1994; Clancy et al, 1999; Clancy and Hansen 1999). The introduction of the Envirochek HV for sampling 1000 L volumes of finished drinking water was an important breakthrough. While the Envirochek HV could easily sample 1000 L of finished drinking water, the elution procedure had to be altered to include a pre-treatment step using sodium hexametaphosphate to permit maximum recovery of oocysts from the sample matrix. In addition, the IDEXX Filta-Max system was also tested and shown to be acceptable for use with 1000L samples. The high volume finished drinking water methods can now be used to develop meaningful data on finished waters, both for IFA analysis as described in Method 1622 or for capture and recovery of oocysts for infectivity testing by any of the cell culture/polymerase chain reaction methods currently coming into use (Rochelle et al, 1997; DiGiovanni et al, 1999).

An alternative method of elution of the Filta-Max foam filters was validated. The method modification using the stomacher/centrifugation technique is an acceptable alternative to the wash station and particle concentrator for elution and concentration of *C. parvum* oocysts from the Filta-Max, and is far less cumbersome to use, both in ease of operation and time-savings. The stomacher technique is an alternative that offers reduced hands-on time to elute the filter and minimizes clean up between samples. Cross contamination issues with the wash station are eliminated with the use of disposable stomacher bags and centrifuge tubes. Many laboratories will find this to be extremely useful, particularly when large numbers of samples require processing under the LT2ESWTR.

One of the problems with Method 1622 was the lack of an alternative product to compete with the Dynal IMS kit. While many vendors had tried to develop IMS kits, none have been commercially successful. Near the end of the project, a kit was developed and commercialized by Aureon Biosystems, which was tested and shown to be equivalent to the Dynal IMS kit for Method 1622, and the adaptations developed in this study. The kit underwent Tier 1 validation for Method 1622 with 50 L source water samples, and was shown to be equivalent to the Dynal IMS kit for use with 1000 L finished drinking water samples.

Another aspect in the method development was to produce methods that, like Method 1622, are simple to use, are reproducible, and offer maximum cost-effectiveness. The new method component(s) are commercially available and meet these criteria.

CHAPTER 5
RECOMMENDATIONS

Water utilities have been monitoring for *Cryptosporidium* oocysts for nearly two decades with methods that were extremely insensitive. The likelihood of detecting oocysts even when hundreds or thousands were present was low. The improvements noted with Method 1622 permit far greater sensitivity than with the ICR method and data from the ICRSS (Connell et al, 2000; USEPA, 2001e) and other studies (LeChevallier et al, 2002) bear this out. Method 1622 and the variations developed in this study still do not provide information on oocyst species or on the viability or infectivity of the oocyst. However, the ability to couple these methods with an infectivity step is at hand.

Cell culture techniques have been developed which allow determination of infectivity of a single oocyst, and as such are amenable to inclusion in monitoring methods (Rochelle et al, 1997; Slifko et al, 1997; di Giovanni et al, 1999). The use of cell culture followed by nucleic acid detection using polymerase chain reaction (CC-PCR) has been used in drinking water methods at the research level, but not yet translated into common use. Di Giovanni et al (1999) have used their method successfully for finished water monitoring as a final step in Method 1622, replacing the microscopic detection assay. Using this methodology for finished water samples showed that of 720 samples analyzed, 19 (2.6%) were positive for infectious *Cryptosporidium parvum* (Aboytes et al, 2000). LeChevallier et al (2002) used the method in a large watershed study of *Cryptosporidium* occurrence and detected infectious *Cryptosporidium parvum* oocysts in 22 of 560 samples or 3.9%. The CC-PCR step goes beyond the detection of oocysts and provides specific information on species and infectivity, and appears ready to be incorporated into these methods for routine use in water analysis. Only when we routinely assess infectivity of recovered oocysts, have accurate reporting of disease in the population, and understand the source of infections, will occurrence data be useful in making public health decisions and assessing true risk.

REFERENCES

Aboytes, R., G. DiGiovanni, F. Abrams, J. Nieroda, N. Shaw, R. Kozik, and M.W. LeChevallier. 2000. Advances in monitoring finished water for *Cryptosporidium parvum*. Proc. AWWA WQTC, Nov 5-9, 2000. Salt Lake City, UT.

Anon. U.K. Water Supply (Water Quality) Regulations 1989, SI 1989 No. 1147. As amended by SI 1989 No. 1384; SI 1991 No. 1837; SI 2790; SI 1996 No. 3001.

Arrowood, M.J. and K. Donaldson. 1996. Improved purification methods for calf-derived *Cryptosporidium parvum* oocysts using discontinuous sucrose and cesium gradients. J. Euk. Microbiol. 43:89S.

Arrowood, M.J. and C.R. Sterling. 1987. Isolation of *Cryptosporidium* oocysts and sporozoites using discontinuous sucrose and isopycnic Percoll gradients. J. Parasitol. 73: 314-319.

Bukhari, Z., R.M. McCuin, C.R. Fricker, and J.L. Clancy. 1998. Immunomagnetic separation (IMS) of *Cryptosporidium parvum* from source water samples of various turbidities. Appl. Environ. Microbiol. 64 (11):4495-4499.

Chappell, C.L., P.C. Okhuysen, C.R. Sterling, and H.L. DuPont. 1996. *Cryptosporidium parvum*: intensity of infection and oocyst excretion patterns in healthy volunteers. J. Inf. Dis. 173:232-236.

Clancy, J.L., W.D. Gollnitz, and Z. Tabib. 1994. Commercial labs: how accurate are they? J. Am. Water Works Assoc., 86:89-97.

Clancy, J.L., Z. Bukhari, R.M. McCuin, Z. Matheson, and C.R. Fricker. 1999. USEPA Method 1622. J. Am. Water Works Assoc., 91:9 60-68.

Clancy, J.L. and J. Hansen. 1999. Uses of protozoan monitoring data. 1999. J. Am. Water Works Assoc. 91(5):51-65.

Clancy, J.L., Z. Bukhari, R.M. McCuin, T.M. Hargy, C.R. Fricker, Z. Matheson, and N. Sykes. 2000. New approaches for isolation of *Cryptosporidium* and *Giardia*. Denver, Colo.: AwwaRF and AWWA.

Connell, K., C.C Rodgers, H.L. Shank-Givens, J. Scheller, M.L. Pope, and K. Miller. 2000. Building a better protozoan data set. J. Am. Water Works Assoc. 92:10:30-43.

DiGiovanni, G. D., F.H. Hashemi, N.J. Shaw, F.A. Abrams, M.W. LeChevallier, and M. Abbaszadegan. 1999. Detection of infectious *Cryptosporidium parvum* oocysts in surface and filter backwash water samples by immunomagnetic separation and cell culture-PCR. J. Appl. Environ. Microbiol. 65 (8):3427-3432.

Dupont, H.L., C.L. Chappell, C.R Sterling, P.C. Okhuysen, J.B. Rose, and W. Jakubowski. 1995. The infectivity of *Cryptosporidium parvum* in healthy volunteers. N. Engl. J. Med. 332:885-859.

Finch, G. R., C.W. Daniels, E.K. Black, F.W. Schaefer III, and M. Belosevic. 1993. Dose response of *Cryptosporidium parvum* in outbred neonatal CD-1 mice. Appl. Environ. Microbiol. 59:11 3661-3665.

Fireman, M. and R.F. Reitemeier. 1944. Prevention of calcium carbonate precipitation in soil solutions and waters by sodium hexametaphosphate. Soil Science 2:35-41.

Kasper, D.R. 1993. Pre- and post- treatment processes for membrane water treatment systems. Proc. AWWA Membrane Technology Conference, p. 105. August 1-4, Baltimore, MD.

Korich, D.G., J.R. Mead, M.S. Madore, N.A., Sinclair, and C.R. Sterling. 1990. Effects of ozone, chlorine dioxide, chlorine, and monochloramine on *Cryptosporidium parvum* oocyst viability. Appl. Environ. Microbiol. 56:1423-1428.

LeChevallier, M.W., G. Di Giovanni, J.L. Clancy, Z. Bukhari, S. Bukhari, J.S. Rosen, J. Sobrinho, and M.M. Frey. 2002. Source water assessment: variability of pathogen concentrations. Denver, Colo.: AwwaRF and AWWA.

Korich, D.G., M.M. Marshall, H.V. Smith, J.O'Grady, Z. Bukhari, C.R. Fricker, J. Rosen, and J.L.Clancy. 2000. Inter-laboratory comparison of the CD-1 neonatal mouse dose response model for *Cryptosporidium parvum* oocysts. J. Euk. Microbiol. 47(3):294-298.

Matheson, Z., T.M. Hargy, R.M. McCuin, J.L. Clancy, and C.R. Fricker. 1998. An evaluation of the Gelman Envirochek® capsule for the simultaneous concentration of *Cryptosporidium* and *Giardia* from water. J. Appl. Microbiol. 85: 755-761.

McCuin, R.M., T.M. Hargy, J.E. Amburgey, and J.L. Clancy. 2001. Improving Methods for Isolation of *Cryptosporidium* Oocysts and *Giardia* Cysts from Source and Finished Waters. Proc. AWWA WQTC. Nashville, TN, Nov.9-14, 2001.

Okhuysen, P.C., C.L. Chappell, C.R. Sterling, W. Jakubowski, and H.L. DuPont. 1998. Susceptibility and serologic response of healthy adults to reinfection with *Cryptosporidium parvum*. Inf. Immun. 66:441-443.

Okhuysen, P.C., C.L. Chappell, J.H. Crabb, C.R. Sterling and H. DuPont. 1999. Virulence of three distinct *Cryptosporidium parvum* isolates for healthy adults. J. Infect Dis. 180(4):1275-81.

Ortega, Y.R., R.R. Sheehy, V.A. Cama, K.K. Oishi, and C.R. Sterling. 1991. Restriction fragment length polymorphism analysis of *Cryptosporidium parvum* isolates of bovine and human origin. J. Protozool. 38:40S-41S.

Rochelle, P.A., D.M. Ferguson, T.J. Handojo, R. DeLeon, M.H. Stewart, and R.L. Wolfe. 1997. An assay combining cell culture with reverse transcriptase PCR to detect and determine the infectivity of waterborne *Cryptosporidium parvum.* J. Appl. Environ. Microbiol. 63: 2029-2037.

Rochelle, P.A., R. DeLeon, A. Johnson, M.H. Stewart, and R.L. Wolfe. 1999. Evaluation of immunomagnetic separation for recovery of infectious Cryptosporidium parvum oocysts from environmental samples. Appl. Environ. Microbiol. 65: 841-845.

Schock, M.R. 1990. Internal Corrosion and Deposition Control. Ch. 17 in Water Quality and Treatment, A Handbook of Community Water Supplies. F. Pontius, (ed.) Am. Water Works Assn.

Simmons, III, O.D., M.D. Sobsey, F.W. Schaefer III, D.S. Francy, R.A. Nally, and C.D. Heaney. 2001. Evaluation of USEPA Method 1622 for detection of *Cryptosporidium* oocysts in stream waters. J. Am. Water Works Assoc. 93:1 78-87.

Slifko, T.R., D. Friedman, J.B. Rose, and W. Jakubowski. 1997. An *in vitro* method for detecting infectious *Cryptosporidium* oocysts with cell culture. Appl. Environ. Microbiol. 63: 3669-3675.

Stookey, G.E., J.M. Warrick, and L.L. Miller. 1995. Sodium hexametaphosphate reduces calculus formation in dogs. Am. J. Vet. Res. 56: 913-918.

USEPA. 1996. Guidelines and format for methods to be proposed at 40 CFR Part 136 or Part 141. EPA-821-B-96-003. USEPA Office of Water Engineering & Analysis Division (4303), Washington, DC.

USEPA. 1997. 40 CFR Parts 136 and 141. EPA Guide to method flexibility and approval of EPA water methods, EPA Proposed Rule. *Fed. Reg.* 62:14975.

USEPA. 2001a. USEPA Method 1622: *Cryptosporidium* in water by filtration/IMS/FA. U. S. Environmental Protection Agency. Office of Water, Washington, DC EPA 821-R-01-026.

USEPA. 2001b. USEPA Method 1623: *Cryptosporidium* and *Giardia* in Water by Filtration/IMS/FA. U.S. Environmental Protection Agency. Office of Water, Washington, DC. EPA 821-R-01-025.

USEPA. 2001c. Implementation and results of the Information Collection Rule supplemental surveys. U. S. Environmental Protection Agency. Office of Water, Washington, DC. EPA 815-R-01-003.

USEPA. 2001d. Interlaboratory validation study results for *Cryptosporidium* precision and recovery for U.S. EPA Method 1622. U.S. Environmental Protection Agency. Office of Water, Washington, DC. EPA 821-R-01-027.

USEPA. 2001e. Results of the interlaboratory method validation study for determination of *Cryptosporidium* and *Giardia* using U.S. EPA Method 1623. U.S. Environmental Protection Agency. Office of Water, Washington, DC. EPA 821-R-01-028.

LIST OF ABBREVIATIONS

°C = Degrees Celsius

μg = micrograms
μL = microliter
μm = micrometer

AwwaRF = Awwa Research Foundation

CC-PCR = cell culture-polymerase chain reaction
CEC = Clancy Environmental Consultants, Inc.
CFR = Code of Federal Regulations
cm = centimeter
C. parvum = *Cryptosporidium parvum*

DABCO = 1,4- diazabicyclo [2.2.2] octane
DAPI = *4'-6'-diamidino-2-phenylindole*
DI = deionized water
DIC = differential interference contrast optics

FITC = fluorescein isothiocyanate

h = hour(s)

ICR = Information Collection Rule
ICRSS = ICR Supplemental Surveys
ID = identification
IMS = immunomagnetic separation
IPR = initial precision and recovery

L = liter(s)
L/min = liter per minute
LT2ESWTR = Long Term 2 Enhanced Surface Water Treatment Rule

M = molar
min = minute(s)
mL = milliliter
mm = millimeter
mAb = monoclonal antibody
mM = millimolar
MS = matrix spikes
MS/MSD = matrix spike/matrix spike duplicates

N = normal
NA = Not Applicable

nm = nanometer
NTU= nephelometric turbidity units

OPR = ongoing precision and recovery

PBMS = performance based measurements system
PBS = phosphate buffered saline
PBST = phosphate buffered saline with Tween
PCTE = polycarbonate track etch
psi = pounds per square inch

QA/QC = quality assurance/quality control

RPD = Relative Percent Difference
RSD = Relative Standard Deviation

SD = standard deviation
sec = second(s)

U = units
U.K. = United Kingdom
U.S. = United States
USEPA = United States Environmental Protection Agency

x g = times gravitational force

IWA Publishing's authorised EU representative for General Product Safety Regulations is Diane D'Arras, 15 rue Duret, 75116 Paris, France, e-mail: safety@iwap.co.uk.

Printed and bound by CPI Group (UK) Ltd, Croydon, CR0 4YY
18/09/2026
02218938-0007